BACKYARD BEEKEEPING

BACKYARD BEEKEEPING

Everything You Need to Know to Start Your First Hive

DAVID AND SHERI BURNS

Illustrations by Kate Wong

ROCKRIDGE
PRESS

Interior Designer: John Clifford
Cover Designer: Joshua Moore
Art Producer: Tom Hood
Editor: Brian Sweeting
Production Editor: Andrew Yackira
Illustrations © Kate Wong, 2020;
Photographs Girts Ragelis/Shutterstock
p. 49; ChloeChiocchi/iStock, p. 51; Gareth Webb/iStock, p. 68; Kosolovskyy/iStock, p. 81; Kerkez/iStock, p. 82; Olga Ivanova V/Shutterstock, p. 90; northlightimages/iStock, p. 101; nata-lunata/Shutterstock, p. 110; Atlas Studio/Alamy Stock Photo, p. 115; Daniel Prudek/Shutterstock p. 148.

ISBN: Print 978-1-64739-514-8
eBook 978-1-64739-515-5
R0

To Jennifer, Jill, David, Karee, Seth, and Christian, our brood who made our hive into a home and who helped make our dream a reality.

CONTENTS

INTRODUCTION

Our names are David and Sheri Burns and beekeeping is our love, passion, and livelihood. We have been beekeeping for over 25 years, and we enjoy helping others experience the joys of beekeeping for themselves.

Our beekeeping hobby began a bit unexpectedly. Our friend was a beekeeper and asked if we could help with a project. A large tree had fallen over and a bee **colony** inside it needed to be removed. David agreed, our friend suited him up, and they headed for the fallen tree, which was full of over 60,000 **swarming** bees. It was a beautiful afternoon and the bees were going about their normal activities, flying around near the opening in the fallen tree. Our friend informed David that if he saw him run away, he should follow him. Our friend cranked up his chainsaw and began removing the outer layer of the tree to expose the **comb** so they could remove the nest and place it into a nicely prepared **hive**. However, within minutes, the noise of the chainsaw drew every bee out of the tree to attack them. Our friend turned off the chainsaw and began running toward some underbrush. David immediately caught up, then passed him.

The work continued on this tree for several more days, and eventually they successfully moved the comb—hive, **queen**, and all. Of course, this was a rough way to start beekeeping, and we wouldn't recommend anyone begin this way. There are much calmer ways to start this hobby than cutting a hive out of a tree.

Our first year was all new to us. We were fearful of the bees and unsure of our own abilities. When we started in 1994, there were very few classes or books compared to the resources available today—we couldn't watch YouTube videos or do our own online research—and so it took us much longer to learn the basics than it would today.

Every year we kept asking questions and experimenting until we reached a point where we felt confident in our beekeeping knowledge. Although it did not come quickly or easily, the work was very rewarding. At first, it was very challenging performing a hive inspection. After many inspections, we began understanding more about the bees and how they must be handled. We quickly learned that bees must be handled with great gentleness. The steep learning curve was well worth it when we harvested our honey for the first time. Watching the beautiful, amber-colored honey flow into our jars was so satisfying. Even observing the bees flowing in and out of the hive, tending to their work day after day, is a relaxing experience.

Beekeeping is a hobby that requires good old-fashioned persistence and endurance. There will be challenging moments when you cannot find your queen and you wonder if you even have a queen. Some inspections are so fun and enjoyable while others can be disrupted by rain or your **smoker** running out of fuel. Particularly in an urban or suburban setting, challenges may include limited space, navigating local regulations, and the concerns of neighbors. But the rewards greatly outweigh the challenges. These rewards include increased **pollination** in your neighborhood's gardens, raw honey from floral sources in your area, and the knowledge that you're supporting creatures that are essential to a healthy local ecosystem.

We love our hobby of beekeeping so much, we hope you'll find the same rewards and joys in it that we do. Maybe beekeeping is something you've always wanted to do but due to a lack of confidence or time constraints you've been putting it off. We will help build your confidence and guide you through the decision-making process so that you can pursue your dream of keeping bees.

We encourage you to embrace beekeeping as a lifelong learning opportunity. Never give up. Keep learning and, most of all, enjoy it.

PART 1

Planning Your Colony

Prepare for your life to be transformed by honeybees. We've found that beekeeping offers many benefits in addition to honey. Not only will you find yourself out in the fresh air enjoying the beauty, sounds, and smells of nature, but beekeeping is a great way to take your mind off worries and anxieties. You may even make new friends and acquaintances as you participate in local and state beekeeping clubs.

That's the beauty of beekeeping; you can keep your hobby small and private, or you can go big, start a business, make extra income, and finally pursue a lifelong dream. Just remember you have the freedom to make beekeeping what you want it to be, to fit your lifestyle and your schedule.

BEING A BACKYARD BEEKEEPER

Your motivation to pursue beekeeping may stem from a desire for your very own raw and natural honey. Maybe you'd like for honeybees to pollinate your garden, or maybe you simply want to enjoy watching these fascinating creatures go about their daily tasks. Backyard beekeeping is also a great way to maximize your outdoor space and learn more about your local climate and ecosystem.

Depending on your situation, the time commitment can be minimal or extensive. Two 15-minute inspections per month is the minimum, but many beekeepers enjoy working with their bees more often.

As a beginner, the first step is to gain some basic knowledge of beekeeping, such as taking a class or reading this book. Approach beekeeping as an enjoyable and exciting adventure. Take time to walk your yard and plan the best location to place your hive. Be creative and express your inner artist and paint some flowers on your hive.

WHY KEEP BEES IN YOUR BACKYARD?

We spend time every day watching our bees fly in and out of their hives, working so hard yet so effortlessly. The sound of their buzzing and the sight of their complex flight navigations amazes us. How can such a tiny insect, especially one that can sting, play such an important role in society? It is because plants and honeybees share a mutual and codependent relationship. There is great joy in knowing that honeybees play a vital role in pollinating our fruits and vegetables. Without honeybees, our diet would be very bland. As beekeepers, we are helping bees thrive so they can better pollinate plants and trees. As an added reward, bees collect the **nectar** from flowers and make liquid gold: honey.

THE BENEFITS OF HONEY

Years ago, my dentist enjoyed buying honey from us because his dad was diabetic and could tolerate honey better than other sugar. Honey is a more natural sweetener than table sugar. The joy of spreading raw honey on toast, adding it to coffee or tea, or taking it for a sore throat reminds us of the important role honeybees play in society. Enjoying raw honey from your local area is believed to help with allergies, because honey contains small traces of **pollen**, which helps build up your resistance to allergies. Honey is also used to treat wounds, burns, and rashes.

Honey also helps control coughing. In one study published by the Mayo Clinic, "honey appeared to be as effective as a common cough suppressant ingredient, dextromethorphan, in typical over-the-counter doses." Since honey is low-cost and widely available, it might be worth a try.

Honey also contains some antioxidants and minerals. Honey has been shown to fight many infections, including E. coli and salmonella.

Is honey better than sugar? Sugar contains 50 percent fructose and 50 percent glucose, and honey contains 40 percent fructose and only 30 percent glucose. The remaining 30 percent of honey consists of minerals, water, pollen, amino acids, enzymes, antioxidants, and vitamins. Regular table sugar is mainly sucrose because glucose and fructose are bonded together. In honey, glucose and fructose are largely independent of each other. Because of the differences between honey and sugar, our bodies handle honey better. Table sugar passes through our stomach due to its composition and is not broken down until it reaches our small intestine. Honey, however, contains various enzymes added by the bees that make it more easily absorbed by our bodies.

THE IMPORTANCE OF POLLINATION

Honeybees are essential in pollinating foods that are good for us, such as fruits and vegetables. Many other insects are pollinators as well, but honeybees are by far our most valuable pollinators, especially for foods like apples, almonds, melons, cranberries, and broccoli.

According to the FDA, honeybee pollination "accounts for about $15 billion in added crop value." Honeybees are like flying dollar bills buzzing over US crops.

Plants rely on honeybees for fertilization. When a bee visits a flower, grains from the stamen, the male reproductive part of the plant, are carried by the bee to another flower and are dropped off at the stigma, the female reproductive organ. When this takes place, a flower is fertilized and will eventually form a fruit. Without honeybees pollinating plants, there would be either no fruit or misshapen fruit. A cucumber, for

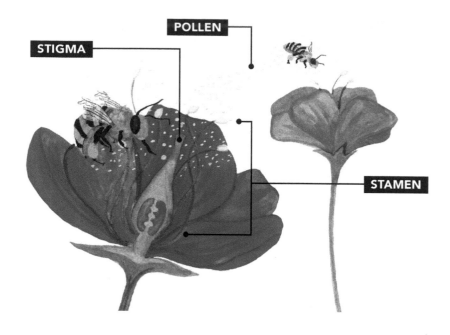

example, might be fully developed on one end from adequate pollination, but underdeveloped at the other end from inadequate pollination.

Flowers attract honeybees with ultraviolet colors that allow bees to identify them more easily. To a honeybee, the colors of flower petals are like landing lights, attracting the bee toward the source of nectar. If directional colors are not enough, flowers give off beautiful scents to attract honeybees as well. Bees are excellent sniffers, even better than dogs, and their keen sense of smell allows them to track down sources of nectar.

SAVE THE BEES

Since honeybees play such an important role in pollination and producing honey, it's no wonder that public interest in honeybee conservation has peaked in recent years. In 2006, the appearance of **colony collapse disorder** (CCD) brought a new level of public awareness to the threats faced by honeybees. Large commercial beekeepers noticed that many of their colonies were gone. They were shocked to find empty hive after empty hive. It was devastating. As the media reported on the decline and disappearance of honeybees, people began worrying about losing the honeybees from the landscape of America.

We had just started our beekeeping business when CCD was identified as a real threat. We thought for sure beekeeping was over. However, immediately our phones began ringing off the hook. People everywhere wanted to start beekeeping to help save the bees. In our home state of Illinois, the number of beekeepers had been steadily declining, but the Illinois Department of Agriculture reported that between 2000 and 2018, the number of beekeepers increased from 1,142 to 4,308. CCD was devastating to honeybees, but it caused beekeeping to surge to record levels and scientists as well as the public sought new ways to save our bees.

A HISTORY OF BACKYARD BEEKEEPING

The earliest hives were nothing more than upside-down baskets covered in mud. Other forms of early hives were tile cylinders and jars. Because honeybees build large comb nests that can nearly fill these enclosures, it was difficult for early beekeepers to harvest honey without disturbing the **brood** area, where the bees raise their young. The first successful attempts at harvesting honey were achieved by placing baskets on top of boxes. Bees would place their extra honey into these baskets, which the beekeeper would then remove. However, beekeepers knew no better way to extract the honey than by crushing it out of the comb, destroying the comb in the process.

In the mid-1800s, beekeeping took a huge step forward thanks to Rev. Lorenzo Langstroth, a brilliant beekeeper. Bees were a comfort to him from his melancholy life as a church pastor. His time spent watching and studying bees led him to design a new hive utilizing what was known as "bee space" in the summer of 1851. He found that by allowing about a quarter of an inch of space around **frames** of comb, the bees would not attach the comb to the walls of the hive or to other frames. This allowed Langstroth to temporarily remove frames of comb, inspect or harvest honey, then place them back into the hive without disturbing the brood area or crushing comb.

WHAT SHOULD YOU CONSIDER?

Starting as a new beekeeper presents several challenges, but these can easily be overcome by properly researching and preparing for your new endeavor. Taking a class is a must. Enjoying a class taught by a certified master beekeeper will give you information that will save you both time and money. Continuing to learn more about bees and monitoring your hives will ensure you are able to address challenges that arise along the way. Like any hobby, there is great pleasure in learning, experimenting, and sharing failures and successes with other beekeepers at local clubs. It is rewarding to know that we can help these great pollinators through our diligent care and stewardship. Proper planning and preparation will certainly curtail many common mistakes.

Many may wonder about the time, space, and personal abilities required to keep bees. Some wonder if beekeeping may be too time-consuming. However, beekeeping can fit into most schedules and budgets. Unlike caring for a dog or cat, beekeeping allows for much more flexibility.

EXPENSES

Like most hobbies, beekeeping can be approached with the attitude of spending the least amount of money or by only buying the best. For example, some people may attempt to build their own equipment to save money and others will purchase assembled and painted hives ready to be placed in the yard. Let's look at the basic equipment and expenses you will need to start. The basic hive costs about $300. A bee suit, smoker, and **hive tool** will cost just under $100. A package of bees with a mated queen will cost around $130. In a perfect year, a beekeeper may harvest 70 pounds of honey and sell it for $10 a pound, which covers the initial investment.

TIME COMMITMENT

Beekeeping does require some investment of time each season, with winter requiring the least amount of time as bees are quiet and clustered together in colder weather. Spring is the busiest season for the beekeeper, as this is when bees begin to grow in population and require more frames to expand. The beekeeper must stay ahead of this rapid growth. Summer requires adding honey **supers** as needed. A honey super is a medium-size box containing frames for bees to store honey. Checking for any pests and diseases is also part of summer inspections. Time is spent in late summer and fall harvesting honey and preparing the hive for winter. Even though bees require very little attention in the winter, this is when most beekeepers spend time preparing equipment for the next year and studying and learning new approaches for improved management.

SPACE

One of the advantages of beekeeping is that it requires very little space. Urban beekeepers with limited yard space often place hives on rooftops and balconies. We are commonly asked how many hives one acre can support. Since bees typically fly two to three miles to forage, an acre can support many hives provided there are sufficient nectar sources within a three-mile radius. The footprint of a traditional hive is around 20 inches by 24 inches. When planning your space, consider leaving enough room to mow around your hives and placing them in an easily accessible place away from foot traffic.

HEAVY LIFTING

Beekeeping does require a certain degree of tenacity. Bees do not always do what we want them to do. However, in general, you can quickly adapt to their various oddities. Others wonder if they are physically able to keep bees. There is a fair amount of lifting. Harvesting honey may require lifting a 40-pound super filled with honey. However, for years we helped

Sheri's father harvest his honey. You can always ask a friend or fellow beekeeper to help you with any heavy lifting.

LEGALITY/RESTRICTIONS

Before committing to beekeeping, check with your state or local municipality. It may be necessary to secure a permit to keep bees. In our home state of Illinois, we are required to register our hives with the Illinois Department of Agriculture. There is no cost, and by registering, beekeepers benefit from having trained inspectors inspect their hives for contagious diseases. Inspectors can also be called to give advice when the beekeeper suspects problems. Many states have similar programs to safeguard the spread of dangerous pathogens. Some states require that colonies be inspected prior to crossing state lines. Contagious diseases are usually found in the brood comb, so inspectors will want to give a clean bill of health before you move your hives to a new state. Your local community may require that you take a beekeeping course prior to keeping bees in the city. When checking with your municipality, read the actual laws for yourself rather than taking someone's word. We have always found state inspectors to be very cooperative and supportive. Having a fresh set of eyes helping you in your beekeeping endeavor can be very helpful (and usually your tax dollars are already paying for it).

REGIONAL CONSIDERATIONS

New beekeepers are often unaware of regional factors, such as nectar sources, weather, winter survival, or other beekeepers in the area. Identifying these influences on beekeeping in your area can help you plan for contingencies and mitigate risks. Take time to seek out local beekeepers who are knowledgeable about factors affecting beekeeping in your area, such as when certain flowers bloom, how changing seasons affect bees, and other environmental challenges. For example, several southern states have unique crops such as orange blossom and tupelo. Sourwood trees, which produce sourwood honey, are mainly found in Tennessee and Georgia at elevations above 1,000 feet. Canada is known for its abundance of clover honey. Knowing about your local climate and plant life can affect your decisions about where to place hives and inform your expectations about what types of honey your bees will eventually produce.

Understanding the length of your bee season will help you know when to work to prevent swarms, split your hive, harvest honey, treat for pests, and prepare for winter. Most new beekeepers are unaware of specific regional nuisances and spend several years learning as they go. Finding out this information at the outset will save you time and effort later in your hobby.

URBAN BEEKEEPING

Keeping bees in the city can be as enjoyable and rewarding as in the country; however, there are several notable differences. In the city, neighbors must be considered, whereas in the country rarely are neighbors close enough to be concerned about your hives. Cities often have specific codes and guidelines for beekeepers to follow, such as requiring beekeepers to take a class or only have a certain number of bees per resident. In the country, most property is considered agricultural and there are no zoning restrictions for beekeeping.

The placement of your city hive must take into account how the bees will exit the hive. Facing the hive entrance toward a wall or fence will force the bees to gain altitude faster and avoid running into neighbors. Place your hives away from your neighbor's pool or play sets, maybe toward the back of your property. Provide fresh water on your property to keep the bees from going to your neighbor's birdbath or pool for water. Also, bees defecate upon takeoff, so do not point the hive entrance toward your neighbor's clothesline or driveway.

Some urban lots may be small, so much care must be taken in placing the hive to minimize exposure to neighbors. Hives can be placed on cement or asphalt driveways, but monitor how hot these surfaces may become in the summer. In the fall, during a nectar **dearth**, bees will be lured into trash cans and dumpsters looking for sweets. Feeding your bees during these dearths will keep your bees from dumpster diving. Finally, controlling swarming will keep your bees from landing on your neighbor's home or trees. Control swarming by making splits or by giving your bees plenty of room in the hive to expand to additional hive boxes.

ALTERNATIVES TO THE BACKYARD

Urban beekeepers often have little to no yard, which simply means it's time to be creative in placing your hive. Hives are often placed on balconies or rooftops or suspended on platforms or hung from trees. One year, we placed several hives on the roof of one of our buildings. The roof had a minimal pitch, so it was easy to work on. The metal roof was not too hot for bees, and we made sure to place a cinder block on the hive to prevent it from being blown off the roof during a thunderstorm. We've even placed a hive on the top of another hive to conserve space.

LOCAL BEEKEEPER ASSOCIATIONS

One of the joys of beekeeping is that you do not have to do it alone or without help. You can join your local beekeeping club and quickly enjoy the camaraderie of other beekeepers. Talking with other beekeepers at your club will answer many of your questions as well as provide you with tips and resources. You can ask fellow members what they do to prepare their bees for winter or if there are specific zoning restrictions or health department regulations for harvesting honey. Local clubs usually pull together to purchase bees and equipment in bulk for discounts. At local meetings, beekeepers share what diseases and pests they are struggling with and what treatment approaches have been successful. Established beekeepers can also give you pointers on how and where to market your honey. If you need a mentor, perhaps you can develop that relationship with someone from the local bee club. Our local club meets at a different member's home each time and opens a hive during good weather.

THE CONSIDERATE BEEKEEPER

As beekeepers, we are ambassadors for the honeybee. In working with our neighbors, we have the opportunity to educate and inform them of the importance of honeybees and not to fear beekeeping. When talking with your neighbors, giving them a jar of honey can sweeten the conversation. Reassure your neighbors that feral bees are already flying around and visiting their flowers and that your bees will probably fly away to larger fields or trees. Never open or inspect your hive when the neighbors are having a cookout or their children are playing outside. Practice swarm management techniques to reduce the chances of your hive swarming and landing on your neighbor's tree. In the fall, during a nectar dearth, help your neighbors keep their sugary trash well covered.

COMMUNICATING WITH NEIGHBORS

Before we were beekeepers, we discovered our elderly next-door neighbor had a backyard beehive one sunny spring day when a swarm of honeybees swooped down on our play set and flew within inches of our children. We were surprised, and more than a little alarmed, until our neighbor was able to calm our fears by answering our questions.

It is very important for a beekeeper to inform their neighbors of their beehives. Some beekeepers will go to all extremes, like our former neighbor did, to hide beehives from the community, but it is far better to establish connections and communicate with your family, friends, and neighbors about the presence of honeybee hives. Your neighbors may be concerned about stings (and possible allergic reactions) or whether the bees will be a nuisance in their pool, pet water dishes, birdbaths, or refuse

SWEETEN THE DEAL

Q: Is it legal to have bees in this neighborhood?
A: Check regulations at your local government office.

Q: Are you trained to do this?
A: Beekeepers take workshops and attend bee club meetings to learn advanced techniques.

Q: Will your bees get in my yard?
A: Most likely, along with feral (wild) bees and other insects.

Q: Do your bees sting?
A: Bees only sting when they feel threatened, so if you stay away from bees, they will stay away from you.

Q: What do I do if I'm stung?
A: Remove the stinger with the sharp edge of your nail or with a credit card. Apply ice or analgesic. Watch for any possible allergic reaction.

Q: What does an allergic reaction look like?
A: An anaphylactic shock is rare and characterized by rapid breathing, wheezing, flushing, fainting, and shock. More common is a local reaction with itching and swelling.

Q: Why are there bees in my trash?
A: Bees are attracted to sugary smells. Keep your trash sealed in trash cans with lids, and clean your trash cans regularly with soapy water.

Q: Bees are in my garden!
A: Work calmly and slowly, and the bees won't bother you.

Q: Why are the bees in my hummingbird feeder?
A: Bees are attracted to the sugary feeder solution. Honeybee-proof animal and bird feeders are now available on the market.

Q: We're having an outdoor party. Will the bees bother us?
A: Bait the bees far from the party with a sugar water solution.

and compost. They may even wonder if you plan to sell your honey, causing possible traffic issues.

These are all valid concerns, and by being an ambassador for the bees, you can relieve your neighbors' fears by answering their questions, offering reading materials, or by hosting a tour or demonstration. Be positive and enthusiastic, giving them your phone number or e-mail address along with a gift of honey or a handmade candle to remind them of the wonderful benefits that bees provide us. Check in with your neighbors periodically for any continued concerns or new issues.

BEEKEEPING APPROACHES

Each beekeeper may approach beekeeping differently. Some feel that beekeeping should be completely natural. Others may only want purely organic honey. Still others may be comfortable using legal medication to improve their hive's overall health and productivity. Each has its share of pros and cons. Most beekeepers incorporate some portion of each of these in their beekeeping practices—it comes down to their philosophy on beekeeping. Large, commercial beekeepers are more likely to use medication, but many backyard beekeepers want to be as natural as possible and only use medication when their bees may perish without it.

MEDICATED BEEKEEPING

Should beekeeping include medicating honeybees? Hives can sometimes receive medication, such as antibiotics. Other forms of chemicals are used to control pests. There are several contagious diseases that can be controlled with the use of antibiotics. State inspectors may require the application of antibiotics when certain diseases are confirmed. As of January 1, 2017, antibiotics to be used in beehives must be prescribed by a veterinarian. This is an attempt by the FDA to slow antibiotic resistance. For example, European foulbrood (*Melissococcus plutonius*) is a bacterial disease that affects developing **larvae** and can be treated with an antibiotic.

Many types of chemicals are available to treat for mites. Some bee-keepers use "soft chemicals" such as formic acid or oxalic acid to help control mites. These medications and chemicals have specific label regulations stating if they can be administered with or without the honey supers on the hive; the label must always be followed.

NATURAL BEEKEEPING

Natural beekeeping means different things to different beekeepers. Everyone would love to be as natural as possible when it comes to keeping bees. Placing a package of bees in a hive, watching them grow, then harvesting the honey in the fall is as natural as it gets. This might work in a perfect world, but in 1987, beekeeping in the United States fell victim to a parasite known as *Varroa destructor*, a tiny mite that attacked hives throughout the country. The mite spreads viruses that weaken and kill hives. Natural beekeepers quickly went to work to find more natural ways to control mites, such as screen bottom boards, powdered sugar dustings, green drone comb trapping, and more. The challenging aspect of natural beekeeping is addressing mites and diseases without the use of chemicals or medications.

ORGANIC BEEKEEPING

Is organic beekeeping possible? Typically, natural beekeeping means that a beekeeper is attempting to keep bees without the use of harsh chemicals or medication. Organic beekeeping is much more challenging, because bees fly miles away gathering nectar. Often the beekeeper cannot know for certain where bees are gathering nectar and if the area only contains organic flowers or crops. This poses a very challenging dilemma. To be truly organic honey, the nectar source would be required to be grown organically as well. Since bees fly three or more miles to gather nectar, this would require that every plant within three to five miles be organic. In most areas, this would be impossible to control. One home-owner spraying their apple trees with pesticides within a three-mile radius would negate the possibility of organic beekeeping. Unfortunately, in the United States there are very few regulations or standard guidelines for labeling honey as organic.

COMBINED APPROACHES

Medicated, natural, and organic approaches to beekeeping all have their place and following. But what is probably most common is a combination of all or some of these approaches. Medicating bees leads to bees having fewer natural immunities. Remaining strictly natural causes bees to suffer from pests and diseases with less help from humans. Organic sounds perfect, but is usually impossible for the average beekeeper. Therefore, a combination of these approaches can be pursued so that the beekeeper helps the bees, bees can become stronger, and yet the colony and their products are as pure as possible given the challenges.

When mites first weakened our colonies, we treated them with legal chemicals. However, these chemicals were soon found to be absorbed into the wax, the incubator and maternity ward of the hive. Over several years, beekeepers quickly began figuring out how to get off the chemical treadmill. Today, we practice natural beekeeping with an occasional treatment if absolutely necessary.

BETTER KNOW A BEE

It is fascinating to observe a colony of honeybees, some 40,000 bees, working together as a single organism. It is important for the beekeeper to learn to think like a bee and to understand the colony's role and function even down to the individual bee. Beekeepers must know bees to keep bees. Without properly understanding honeybees, the beekeeper cannot fully care for and tend to the colony.

A colony of honeybees is made of three types of bees: the queen, drones, and **workers**. The queen is the only reproductive bee in the colony, laying 1,000 to 3,000 **eggs** every day. She also spreads a **pheromone** known as QMP, queen mandibular pheromone. The queen's pheromone is the glue that holds the colony together, and her main role is to lay eggs and spread pheromones. The queen has a circle, a retinue of attendants who feed and groom her as she goes about her tasks. The drones are the male bees. They do not forage for resources, protect the hive, or do much of any hive maintenance. Their only role is to mate with new, virgin queens from other colonies, then they die. Finally, worker bees are the workforce of the colony. The workers are all female workers who perform various duties until they reach the age of 23 days, at which time they began foraging for resources. For example, a **forager** will continue to forage for water, nectar, and pollen. Pollen carries the male gametes of seed plants and so as bees travel from flower to flower, they are fertilizing plants. A forager also gathers **propolis**, a sticky substance gathered from plants and trees that is used to strengthen the comb, cement the comb in the hive, and control pathogens. Bees forage until they reach the summer age of 40 to 45 days old. This is the age at which a summer bee dies from having worked herself to death.

THE HIVE

Various types of bees build different nests. For example, mason bees use mud to construct their nests or will nest in natural cavities such as hollow plant stems or holes made by other insects. Many bees nest in the ground, such as ground bees, bumblebees, and yellow jackets.

Honeybees build their nest in hollow trees or deserted openings in barns, as they prefer a completely dark enclosure. Beekeepers quickly learned to provide a spacious box to attract them. In beekeeping, a hive refers to the woodenware, and is where a colony makes its hive. Within the cavity of the hive they begin building comb, suspending it from the top of the cavity and building it downward. The orientation of the hive always places the brood near the opening and the stored honeycomb far from the opening. This is to add greater protection for the honey.

SWARMING

Honeybees reproduce by creating a swarm in the spring. A swarm is when 50 percent of the colony leaves with the old queen to find a new nesting site. Prior to the day of the swarm, bees will spend weeks planning for the event. Scout bees will explore new potential homes. As the day draws near, foragers will forage less as they prepare for the getaway event. Even the queen is fed less so that she will be better able to fly with the swarm.

When the swarm first leaves the original hive, they land close by the hive, usually in a tree. Scout bees perform **waggle dances** to lure the swarm to the new nesting site. The waggle dance was first observed by Karl von Frisch. He determined that scout bees would waggle or vibrate their bodies in a straight line to indicate the distance or energy bees must put

into the flight. The angle of the waggle dance respective to the location of the sun indicates the direction of the flight. Other bees observe these signals and know exactly where to fly. At first, each individual scout bee may compete for their site to be chosen, but soon all scout bees will agree on one site. It is at that moment that the swarm will lift off again and head to the new site. The original colony raises several new queens to replace the one that left. A dozen or more new queens emerge in the original colony and fight it out until only one remains to be the new queen.

Swarm prevention is very challenging, as a colony must reproduce. However, beekeepers can divide or split their colony just prior to swarm season in the spring. This artificial swarm can lead the colony to believe they have swarmed.

THE LIFE OF BEES

Understanding the life cycle of honeybees enables beekeepers to better manage and care for their hives. Female worker bees perform myriad tasks according to their age. At first, young bees perform tasks close to the center of the brood nest, such as caring for young larvae. As they age, they venture farther away from the center of the hive to build comb on the edges of the brood nest and transport nectar. Finally, they participate in the riskiest behavior, foraging miles away.

TERMINOLOGY

Egg. Small, white, and shaped like rice; bee eggs are found in the base of a cell of honeycomb.

Larva/Larvae. A legless and featureless white grub.

Pupa/Pupae. The stage of metamorphosis in which the tiny bee begins to develop its adult body parts; a larva enters a cocoon in order to transform into a pupa.

Adult. In the last stage of metamorphosis, the fully formed adult bee emerges.

Brood. A term that refers collectively to the eggs, larvae, and pupae of the honeybees.

Workers. Bees that are responsible for housekeeping, taking care of the queen and brood, and foraging; these bees are always female.

Drones. Male bees without stingers whose only role is to mate with new, virgin queens from other colonies.

Queen. The only egg-laying female in the colony.

Nectar. A sweet liquid produced by flowering plants and collected by honeybees.

Pollen. Powder-like substance produced by the male parts of a flower.

Propolis. A sticky substance gathered from plants and trees that is used to

CONCEPTION

A queen can choose to fertilize an egg or not, which determines whether it will develop into a female bee (fertilized) or male bee (unfertilized). Each cell of the honeycomb is prepared to receive an egg and is carefully cleaned and polished by house bees. Once prepared, the queen will examine the cell and measure it with her front legs. If the cell measures to be a worker-sized cell and is prepared properly, she will walk forward, then back her abdomen into the cell to deposit a fertilized egg. Drone-sized cells, made for male bees, are slightly larger. This triggers the

strengthen the comb, cement the comb in the hive, and control pathogens.

Royal Jelly. A substance secreted from the hypopharyngeal glands of nurse bees and fed to young developing larvae.

Bee Bread. A mixture of honey, pollen, nectar, and bee saliva that creates a stored, fermenting food source for the colony.

Absconding. When the entire colony abandons the hive; this can happen for a variety of different reasons.

Dearth. A condition in which there are few to no available floral sources for bees to forage.

Honey Bound. A condition resulting from an abundance of foraging that has filled most frames with nectar or honey, limiting the area for brood or pollen.

Moisture Level. The percentage of moisture in honey. Honey is best harvested when the moisture is around 18 percent.

Refractometer. An instrument used to measure the moisture content of honey.

Queen Excluder. A divider made of material with spacing that allows workers to pass but excludes the queen. It is used below a honey super to prevent the queen from laying eggs in the super.

Laying Workers. In the absences of the queen and/or brood pheromones, the workers begin laying unfertilized eggs that mature into male drones.

queen not to fertilize the egg. As a backyard beekeeper, each biweekly hive inspection will involve checking for eggs and ensuring the queen is laying adequately. A pair of reading glasses, a magnifying glass, or a quick snap of the cell phone camera can help identify the tiny egg in the base of a cell. It looks like a small grain of rice placed perfectly in the middle of a cell in the comb.

METAMORPHOSIS

As the queen lays fertilized eggs, they become female workers or queens. Unfertilized eggs become male drone bees. When the queen lays an egg, she lays it in the base of a cell in the comb. The egg stands straight up on the first day, like a piece of rice on edge. On the second day, the egg begins to lean, and finally on the third day it will lay flat on the bottom of a cell just prior to when it metamorphoses into a larva.

After three days as an egg, it molts (metamorphoses) into a small, grub-looking larva. The larva almost exclusively consumes **royal jelly** for the first three days. Royal jelly is secreted from the glands of nurse bees, young bees between 5 and 12 days in age. After the third day as a larva, the developing bee is fed worker jelly until around day eight, when the larva molts into a **pupa**. Worker bees cap or seal over the top of the pupa cell, and the young pupa begins to spin a cocoon in the cell. In the pupa stage, the developing bee begins to grow, developing visible bee characteristics, and is bright white in color. Finally, worker bees emerge as adults 21 days after the egg was laid. Queens emerge on day 16 and drones emerge 24 days after the eggs were laid.

THE QUEEN

The queen is the all-important, essential, and indispensable bee in the colony. She is vital because she is the only female providing for the future of the colony by laying over 1,500 eggs per day. Without her, the hive would rapidly fall in population and perish in two months. The queen spreads her pheromones, which hold the hive together as a single

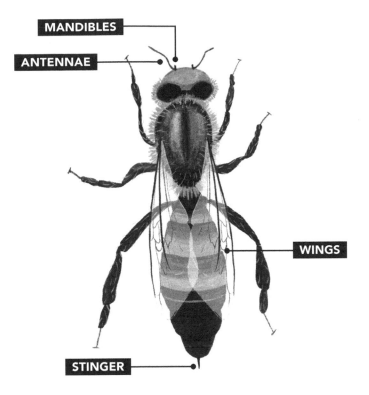

MANDIBLES

ANTENNAE

WINGS

STINGER

QUEEN BEE

superorganism. Monitoring the queen's progress is the backyard beekeeper's primary concern. If the queen fails to lay enough eggs, the colony will suffer from a low workforce.

Many new beekeepers feel pressured to locate their queen to ensure she is present and accounted for. However, it is more important to verify that frames are full of eggs, larvae, and pupae. In our classes, we stress that once you see one-day-old eggs, you need not locate the queen. One-day-old eggs confirm the queen is present and has laid eggs within the last 24 hours. Finding the queen is one of the most challenging aspects of a hive inspection. Many beekeepers become very frustrated. Marking the queen with the appropriate color for that year can help identify her quickly. You can learn how to pick up your queen and place a dab of nontoxic paint on her thorax (between her head and abdomen) so she is more easily spotted. Finding the queen is best accomplished by looking in the brood nest area. The queen is rarely found on the honey frames on the outside of the brood nest frames. Brood nest frames are in the center of the hive, usually the center six frames. Once you locate frames that contain brood, keep looking for younger brood until you see eggs. Once you see eggs, the queen is very close by.

If the queen dies, within 24 hours the colony perceives the loss of her pheromones and quickly begins raising a new queen. A beekeeper can introduce a newly mated queen by purchasing one, keeping her in the cage she's delivered in, and allowing the workers to slowly eat through the candy plug on her cage as they become familiar with her pheromones. If left to raise their own queen, workers begin feeding a young larva royal jelly and drawing out her cell vertically along the outside of the comb. Once this new queen emerges in 16 days, she will walk around the brood nest area for several days as a virgin queen before taking her mating flight. She will take one, maybe two mating flights during the time when she is a virgin. She will fly a mile and a half away to a drone congregation area. Drones from many different colonies, but not hers, are there waiting to mate with virgin queens. She flies into the cloud of male drones and mates with over twenty drones during her 17-minute mating departure from her home hive. She then flies back to her hive with the male genitalia from the last drone still attached (called a "mating sign"). At the moment of her arrival, the future of the colony is now secure.

WORKERS

Population growth ensures a colony's healthy future. A new colony usually reaches 40,000 worker bees by midsummer if started in the spring, which means the hive needs to grow by an enormous amount—1,500 bees per day. Even though this seems like a lot, remember that a bee only lives 40 days in the summer, so the workforce must constantly be replaced.

Bees can be marked with numbers or paint on their thorax and observed to determine their changing role as they age. For example, upon emerging

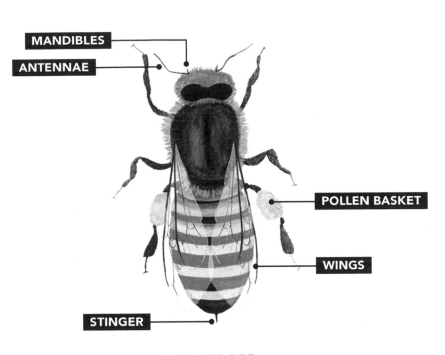

MANDIBLES

ANTENNAE

POLLEN BASKET

WINGS

STINGER

WORKER BEE

from her cell on day one, the new worker bee begins cleaning her own cell and other cells as well as keeping the brood warm. This new adult bee is hairy, very light in color, and soft. A one-day-old bee is so soft it is not able to sting. As she continues to age, on day three to five days old, she will begin providing food for larvae over three days old. Larvae between one and three days old are fed royal jelly. Prior to becoming a nurse bee at five or six days old, young bees cannot produce royal jelly, so they feed older larvae by carrying resources to them. Nurse bees also serve as companions to the queen, known as her retinue. Since the queen consumes mainly royal jelly her entire life, nurse bees continue to attend to the queen.

As the adult worker bee continues to age, she takes on various roles according to the maturity of her glands. For example, between days 12 and 17, the worker bees lose their royal jelly glands and begin using their newly developed wax glands located on the undersides of their abdomens. Now the 12- to 17-day-old adult worker produces wax flakes that are secreted

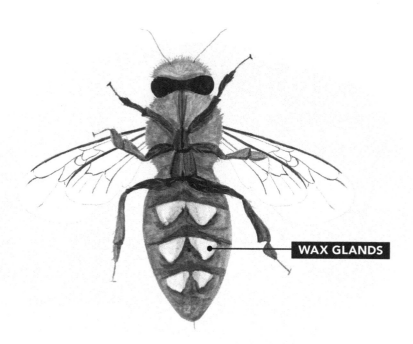

UNDERSIDE OF WORKER BEE

from her wax glands. Worker bees begin building the comb and transporting food at this age. Building combs on frames allows the colony to develop rapidly, as now they have comb in which to place nectar and pollen and for the queen to fill with eggs. During their wax-building days, bees consume 11 pounds of nectar to produce one pound of wax. Bees at this age transport nectar from incoming foragers into honey cells by receiving a droplet of nectar, nearly 80 percent water, from a forager and carrying it in their mouthpiece for twenty minutes, drying it and adding enzymes. The nectar is now placed in cells, and workers fan their wings to evaporate the excess moisture to reduce it to 18 percent moisture, transforming it from nectar into honey. Foragers arrive and carry pollen into the brood nest area, where it is packed into and stored in cells by these house bees.

Around day 18 to 21, the worker bee begins keeping an eye on the hive entrance, guarding against foreign, unwanted insects such as moths, wasps, or robber bees. Finally, around day 23, worker bees begin foraging, which they will do for the rest of their lives (another 18 days). They will forage for nectar, pollen, and water. Water is used to cool the hive and added to the honey when necessary. Worker bees are also collecting propolis, a sticky substance gathered from plants and trees that is used to strengthen the comb, cement the comb in the hive, and control pathogens. Colonies inside trees or hives with rough wood use propolis to smooth the walls, providing health benefits to the colony by controlling decay.

BEE STINGS

The backyard beekeeper cannot keep bees without being aware of the potential of being stung. When someone learns we are beekeepers, their first question is how many times we've been stung. We usually tell them it depends on what specific task we are performing. When removing bees from buildings, there is an increased chance of being stung because we are transferring the bees from their snug home to a new home. But when inspecting our own hives, being stung is rare. There is a good reason that honeybees have stingers. Their hive is filled with a precious treasure that almost all other animals and insects would love to have: honey. However, when a honeybee stings, she dies. Learning to minimize the risk of being stung will make beekeeping much more enjoyable.

First, choose a good day to work your bees, a nice, sunny, warm day when bees are out foraging. Usually the window between 10:00 a.m. and 2:00 p.m. provides the optimal time when foragers are out working. Avoid wearing dark clothing, as brown and black bears are a natural enemy of bees. Brighter colors do not seem to alarm bees, such as white bee suits. Avoid wearing heavy perfume, as the added fragrance may attract bees to examine a smell that mimics flowers. Always work in slow motion. Rapid motions with your hands can cause bees to become attracted to the movement, such as swatting at bees. Avoid blowing on bees or being loud while inspecting your hive.

BUYING BEES

Buying bees sounds simple, but it involves careful consideration. Bees are sold two ways, as **packages** and as **nucs**. A nuc is usually four or five frames of bees, brood, honey, and pollen from an overwintered, strong colony. Packages of bees sell out quickly after the first of the year. Our packages go on sale on January 1, and we are usually sold out around the middle to end of the month. This is unfortunate for beginners who develop an interest in beekeeping in late winter but cannot find any bees available. A package is made up of 10,000 bees and one mated queen. The queen is in her own separate cage. This is known as a three-pound package of bees. The cage is slightly larger than a shoebox and is either plastic or wood and screen.

It is best to locate bees in your area close enough so you can drive and pick them up. Some bee companies ship bees in packages, but you are at the mercy of how well your new bees can survive being shipped.

A nuc contains four or five frames of live bees, frames of eggs, larvae, and pupae, and frames of honey and pollen. The nuc also contains a mated queen that is laying well. Packages are available several weeks before nucs and are less expensive than nucs. However, nucs come with frames that are completely drawn out with comb and filled with brood and other resources, whereas packages are usually placed on new **foundations**, whose undrawn comb will take days and weeks to be fully drawn out.

Consideration must be given to your bee provider. Do they have a good reputation and do their bees perform well? Various types of bees can show desirable traits, but these traits and characteristics are nearly impossible to maintain. Queens mate in the open, making it nearly impossible to hold tight genetic lines. Even if you start out with a very specific type of queen, she will eventually be replaced by your colony, sometimes even in the first year.

We select our stock more on performance than on a particular region or type of queen.

DRONES

Drones are male bees and they make up anywhere between 200 and 2,000 bees in a 40,000 bee colony. Their only role is to mate with virgin queens and then die shortly thereafter. Drones are fed by workers, allowing them to maximize their flight time in drone congregation areas, mating with virgin queens. During their flights, drones are allowed into any hive to refuel and continue their attempt to mate with virgin queens. In our queen-rearing yards, drones can be seen leaving the colonies around mid-morning to mate with virgin queens. Amazingly, by late afternoon, drones that were unable to mate with virgin queens begin returning to their hives in large numbers, for better luck the next day.

The male bees, drones, are slightly larger than the workers and are strong fliers in their ability to mate with virgin queens. Their large compound eyes touch in the middle of their head. Their main role is to mate

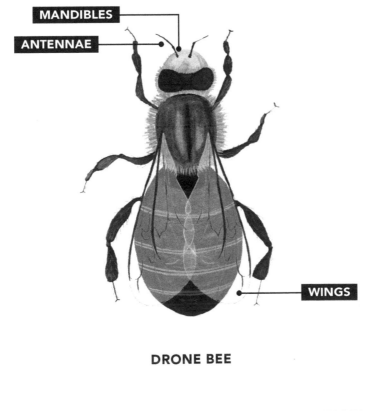

MANDIBLES

ANTENNAE

WINGS

DRONE BEE

with virgin queens, but drones occasionally attack intruders when the hive is being threatened. They have no stingers, only the male genitalia, but can still attempt to inflict fear by bumping the intruder. Drones have smaller honey tanks only used for fueling their mating flights, not foraging. They have no pollen baskets, so they do not gather pollen. Drones have one extra section to their antenna believed to assist in isolating the virgin queen's pheromone during a mating flight. They are very easy to find and distinguish from workers. When we are teaching outdoor courses, especially courses to children, David will place a drone in his mouth, then open it slowly so they can see the drone fly out and away. This appears amazing, because we never tell anyone it is a drone without a stinger. They think it is a worker bee in his mouth. However, before you try this trick, be sure you know the difference between a worker and a drone.

During the spring, drones are extremely important, as this is when colonies swarm and require new virgin queens to mate. Colonies produce an abundance of drones to fill the drone congregation area and ensure new queens can be mated. Drones can fly and mate at about two weeks of age. They are good mating partners until day 28. In queen-rearing operations, hives consisting mostly of drones, known as drone mother colonies, are placed on the outer perimeters of mating yards to ensure virgin queens are well-mated. In more specific queen-breeding programs, specific drones are selected, and their sperm is gathered to artificially inseminate virgin queens in order to raise a particular race of queens.

In colder and seasonal climates, drones are not needed during the winter months. A colony going through winter with a full complement of drones would create a greater demand on food from the colony. In late summer and early fall, colonies are no longer making queens that need to be mated, so drones are kicked out of the hive. Essentially all drones are pushed out of the hive prior to winter except for a few here and there that overwinter, and new drones will be raised in the spring. The ground in front of a hive will be covered with dead drones in the fall.

TYPES OF BEES AND THEIR CHARACTERISTICS

Italian	Gentle, good brood pattern, minimal swarming, good honey production, light propolis.
Caucasian	Longer tongue reaches bottom of flowers more easily. Very gentle. Buildup slower in the spring as they prefer to gather ample nectar and pollen. They also gather more propolis, making the hive very sticky to work. Can rob more.
Carniolan	Explosive spring buildup, are not so prone to rob, very gentle, and good comb producers. Explosive buildup means more swarms in the spring. Honey production can be slightly less than that of the Italian bee.
Russian	Bred to be more mite-resistant and more winter-hardy. Some swarm more.
Buckfast	Originally bred to show a resistance toward tracheal mites and show good hygienic behavior.
Minnesota Hygienic	A result of the work of Dr. Marla Spivak of the University of Minnesota. Bred to reduce mites by being exceptionally hygienic.

CHOOSING THE RIGHT HIVE

omesticating honeybees over the years has inspired many styles and philosophies regarding hive designs. This progress has grown from baskets to boxcs to frames designed to respect bee space, allowing bees to remain a working colony while being inspected and manipulated.

The Langstroth hive was developed by Rev. Lorenzo Langstroth in the mid-1800s. His hive promoted removal frames by providing "bee space," so frames were neither glued with propolis nor attached with wax by the bees. Frames could be easily lifted, inspected, and replaced. The Langstroth hive was the single greatest advancement in beekeeping.

A top bar hive is a horizontal hive, and only the tops of frames are used to promote natural comb growth. The tops of the frames, called bars, are lined up on top of a large trough-like horizontal box in which the bees build their natural comb, attaching it only to the top bars.

The Warré hive is a mixture of a Langstroth and top bar design in that it is a vertical hive like a Langstroth but only uses top bars.

GUIDELINES FOR CHOOSING A HIVE

Basic beekeeping practices must be learned and followed regardless of the type of hive chosen. The backyard beekeeper's skill level and knowledge will bring about a greater level of successful beekeeping than the nuances of a particular hive type. Answering key questions about your beekeeping endeavor can help you decide on the hive that is right for your particular needs, budget, and challenges. For example, will your hive be primarily for pollinating your garden or for bulk honey production? Are you considering a hive for the ease of adding and removing honey supers? Maybe you are wanting a type of hive that is the easiest to build. Will you be moving your hive? Are you wanting a light hive that can be moved, or a heavier hive? Will you be harvesting your honey frames with a traditional **extractor** or crushing the comb to harvest the honey?

LANGSTROTH HIVE

This traditional hive, known as a Langstroth or Lang hive, was first designed by Rev. Lorenzo Langstroth. The Langstroth hive simplified beekeeping by allowing the beekeeper the freedom to lift, remove, and reinsert frames. By far, the Langstroth hive is the most commonly used hive in beekeeping. We started with a Langstroth hive and though we have experimented with other hives, we are by far most familiar with Langs. The beekeeping industry has developed numerous accessories that are all designed to fit the Langstroth hive. Since all manufacturers respect traditional measurements, all parts are interchangeable. Even if you purchase a honey super from a friend, if you both use Langstroth hives, it will fit perfectly.

KEY FEATURES

A Langstroth hive can be configured for either 8 frames or 10 frames, but the most common is 10 frames. Beginners find Langstroth hives easy to assemble and work during their first year. Parts are easy to remove during inspections. The hives are identical except for the number of frames in each box. This gives the 8-frame hive a slightly smaller footprint and reduces the weight by about 30 pounds per hive. A Langstroth hive consists of a bottom board upon which all the other parts rest. The other components include a deep **hive body** consisting of frames of brood comb, sometimes only one in warmer climates and two where winters are harsh. Honey supers are then placed above the brood boxes. These super boxes contain frames of comb where honeybees place excess honey for beekeepers to harvest. The inner cover is placed above the honey supers to provide an air space between the colony and the top cover. The top cover telescopes over the inner cover and an inch or so over the top box to protect the entire hive from wind and rain. Honey production will vary based on the abundance of nectar; however, the average is around 70 pounds per year.

PROS

For beginners, a Langstroth is easy to assemble. When we first started beekeeping, we received our hives in hundreds of parts and pieces with nails and a tiny hammer. We even had to build our own frames and insert the wax foundation. Today, most new beekeepers prefer to begin beekeeping with a fully assembled and painted hive.

Another pro of a Langstroth hive is that it is fully supported by a host of add-ons such as **queen excluders**, feeders, hive stands, scales, propolis and pollen traps, and **entrance reducers**, all designed to fit a Langstroth hive. Also, the vertical nature of a Langstroth hive allows bees to move upward as they would in their natural habitat, such as in a hollow tree. Beekeepers can continue adding additional boxes and allow for the continued growth of the colony.

Honey super frames can be easily removed and harvested using a honey extractor, which is designed to accommodate Langstroth frames. The hive is easily supported on many different types of hive stands, such as cinder blocks, pallets, or homemade or purchased stands.

CONS

I remember being surprised to hear a well-known entomologist (a scientist who studies insects) from a university say that a Langstroth hive leaves much to be desired. He compared a Langstroth hive to the natural honeybee hive in a tree. Immediately we realized that the Langstroth hive is usually more confining than a large hollow tree, especially considering each comb is restricted in size. In nature, a colony can build comb as long or wide as space allows. When we remove hives from between walls in homes, sometimes the comb reaches from the ceiling to the floor in the wall. However, in a Lang, the comb can only be as long and as wide as the individual frame allows.

In nature, a colony in a tree usually has considerable distance from the bottom of the comb to the base of the tree, sometimes several feet. This is beneficial as it allows debris and varroa mites to fall far away from the colony. However, in a Lang, the bottom board is only a few inches from the bottom of the nest area, allowing mites and small hive beetles to regain access to the brood nest area if they fall off. The weight of supers of honey and brood nest boxes can be very challenging, weighing upward of 30 to 80 pounds.

TOP BAR HIVE

Top bar hives can bring a breath of fresh air into beekeeping with a hive that is all horizontal, promotes natural comb, and is fun to build. Many enjoy a top bar hive because of its near waist-high position, no lifting of heavy boxes, and simple manipulation. We made our first "for fun" top bar hive from scraps of wood we had lying around. Once it was built, all we had to do was dump in a package of bees and we were in business. A

national cable news channel asked if they could come out and do a segment on beekeeping. They loved the top bar hive and placed a camera inside it for a live video feed throughout the day.

KEY FEATURES

All Langstroth hives must follow the same measurements, but top bar hives can be built to your personal taste. Top bars are made of two sloping, angled sides with flat end walls. Small pieces of wood similar to the size of the top bars on a Langstroth frame make up the starter strips for bees to build comb. The angle of the sloping walls can be of your choice. The one we built left about a three-inch opening at the bottom where the two sloping sides did not meet on purpose. We then placed a ⅛-inch hardware cloth in this opening for ventilation and for mites to fall out. Then, we made a sloping roof, which worked out nicely. At first, we were concerned that the bees would attach their natural comb to the sides of the walls. In a few places they did connect it with connective wax. We took a long knife and gently cut away these connection points, allowing us to lift the frame out for inspections. The combs were beautifully constructed by the bees, and the hive performed very well. We drilled three small ¾-inch holes on one end of the hive for the entrance, plugging all but one with cork until the colony grew larger and needed more openings.

PROS

A top bar hive has many pros. We enjoyed that the entire hive was at the perfect height for us to work and inspect it. We noticed that as we inspected the hive, each of the top bars remained in place, holding the bees down. An opening only the size of one frame exposed the bees. We enjoyed that we never had to lift off heavy honey supers to inspect the hive. Most top bar beekeepers believe the colony benefits from being able to build natural comb, hanging it from the top bars only. There are no foundation or side or bottom bars to the frame. Therefore, the bees build their cell size without the influence of a foundation that provides a template for size.

Harvesting honey does not require an extractor. Instead, honeycomb is cut from the top bars, crushed, and strained. The honey super wax can be rinsed and melted and reused in the hive or sold. There are no extra honey supers to store until next season. Many backyard beekeepers enjoy this less-expensive approach to starting a hive.

CONS

While top bar hives have their benefits, they also have drawbacks. Once the honey super combs are crushed and strained, the colony must rebuild new comb, which requires significant energy and time. Since unlimited boxes of honey supers cannot be added to a top bar hive, this can result in less honey production. This loss of time will mean a delay in additional honey being stored. The horizontal hive also makes it more difficult to exclude the queen from the honey super combs. Medicating a colony in a top bar hive takes special manipulation.

Most top bar hives are placed on legs about three feet tall. During a harsh winter, the hive is positioned to allow winter winds to flow around the hive. In vertical hives, bees can move up into the honey supers and the warmth of the upper hive. In a top bar hive, a winter cluster must move out of combs to move more into honey supers. We were unable to successfully overwinter our top bar hives after two separate attempts. A top bar hive has a larger footprint, which is something to consider if space is limited.

WARRÉ HIVE

The Warré hive, known as the People's Hive, was developed by Abbé Émile Warré in France in the early 1950s. He experimented with over 350 various designs to finally find a design that is a mixture of a Langstroth vertical hive with frames designed much like that of a top bar hive. The idea is a hive that resembles a Langstroth hive but is designed to be much more natural. Warré hives are more labor-intensive and have a much smaller following.

KEY FEATURES

The Warré hive features boxes slightly smaller than Langstroth size. Most frames are top bars only, although some have side bars to help guide the bees making comb. Warré followers want to provide a much more natural experience for bees, allowing them to build comb at their own cell size preference. Some believe bees prefer to build smaller-size cells, less than 5 mm. Above the brood boxes and honey supers, the Warré hive has what is known as a **quilt box**. This quilt box has burlap or other material as a bottom to the box. The box is usually filled with sawdust or other absorbent material. This provides insulation on hot summer days and helps

absorb winter moisture. Some feel it is important to limit inspections so as not to disturb the natural pheromones used within the colony.

PROS

In the natural habitat of a tree, bees build comb from the top of a hollow part downward. A true Warré approach is to continue to add additional boxes as needed below the current box where the bees are working. This is the opposite of a Langstroth hive, where boxes are placed on top. The idea of adding additional boxes below the current box is to allow bees to continue to build downward, more naturally. By adding boxes from the bottom, a Warré hive promotes not disturbing the hive and maintaining constant pheromone control.

A Warré hive promotes harvesting an occasional honey super from the top; crushing and straining the honey can be accomplished through a cheesecloth. The excess wax can be rinsed and collected or melted into molds for later use. The quilt box is useful to insulate and to wick away excess moisture forming in the hive during winter months.

CONS

A Warré hive is very challenging physically as the hive must be lifted to add additional boxes from the bottom. The hive could possibly weigh close to two hundred pounds. It would be awkward and challenging. A Warré hive is slightly smaller, thus possibly limiting the growth of a colony. This smaller size may also limit the overall production of honey due to a smaller colony population. The comb must be destroyed and cannot be placed back into the hive, like when bees were kept in skeps, which are basically upside-down baskets where early beekeepers would house their bees. When we built our Warré hive and made our quilt box, we were concerned about the quilt box. The burlap is designed to protrude and hang outside the hive to wick away moisture from within the hive. However, in our climate, the burlap was always soaked from winter rain and snow. The smaller footprint and size mean the taller the hive, the more easily it can fall over. Careful attention must be given to keep the hive carefully balanced and weighted from the top.

OTHER HIVES

As beekeeping techniques continue to advance, new hive types are continually being presented. When Rev. Langstroth first developed his hive, it was met with opposition and suspicion. As new hives are presented today, we must remain open-minded and carefully weigh their merits.

FLOW HIVE

The Flow Hive was developed by a father and son from Australia and gained widespread popularity, especially among beginners, because it presented an easier way to harvest honey: just turn a valve. The hive is basically a Langstroth hive, but the honey super, made up of special plastic parts, allows the frames to separate so honey flows out and through a spigot. It still requires the same amount of hive inspections and management. A similar design was patented in 1939, but the patent ran out.

OBSERVATION HIVE

An observation hive is a major draw and attraction, especially for children. An optimal observation hive is not designed to be self-sufficient. Rather, a beekeeper pulls four to eight frames from a strong colony and places them into the glass observation hive. These frames must continue to be exchanged with the mother colony for maximum health. It is important to keep a blanket or cardboard over the glass when the hive is not being observed as young brood must be raised in the dark. Observation hives must be well ventilated and have a tube going through the wall to allow foragers to bring in resources.

PLANNING AND EQUIPMENT

fun part of any hobby is making plans and arrangements to purchase needed items, obtaining equipment and gear, and deciding the best place to get set up. In beekeeping, that means either purchasing or building your hives; collecting your equipment, such as smokers, protective gear, and hive tools; deciding exactly where you'll put your hives; and connecting with a company or friend that can supply your bees.

MAKING SPACE FOR YOUR BEES

We live on four acres in the country on the Illinois prairie. It's not a lot of land considering the average farm here is several hundred to several thousand acres. It always surprises people to find out that we have had nearly 100 hives on our few acres. We always joke about how hard it is to keep our livestock behind the fence and out of the neighbors' fields.

Exactly how much space do you need for bees? The short answer is not a lot. The longer answer is that it depends on several important factors that you will need to consider.

Let's consider first how much space you'll need. When bees leave the hive in search of sustenance, they will fly as many as several miles to return with nectar and pollen, meaning the actual space in your own yard or garden may not factor much. But where will your bees fly to? In a healthy scenario, the bees will have access to a wide variety of floral sources, whether from garden crops, trees, or flowers, along with fresh water. Commercial apiarists will often keep four hives on a single pallet, but backyard beekeepers who choose to have more than one hive will space hives 6 to 10 feet apart from each other to help eliminate drifting of bees from one hive to another, discourage swarming and robbing, and give you the physical space you need to work. Beekeepers generally work colonies from the back of the hive, so make sure you have ample space to walk around the hive, as well as an area where you can stack boxes as you do your inspections. You'll also need some storage space for your equipment and empty supers.

Where to put the hives is a big consideration. Ideally, hives should be in the sun, which warms up the hive, getting the workers out faster while helping to keep some pests and insects out of the hive. Placing the hives east-facing is a bonus as it keeps wind from blowing into an entrance from the other directions. Providing a windbreak either with an outbuilding, bushes, or trees is a plus. A windbreak can alternatively be created with stacked hay bales or a three-walled loafing shed. We have a friend who keeps his hives on a flatbed, which he simply pulls around to the back of the barn when it gets cold and windy outside.

What if you live in a small space? Get creative! You can find hives on apartment balconies and in courtyards. Restaurants tuck hives into tiny herb gardens, and small-town beekeepers may choose to keep them on a shed roof. Some artists have designed innovative hives that resemble small picture frames, which can be hung on walls and have tubes to the outside that allow the bees to come and go. A friend who has an RV keeps his hive strapped on the back of the vehicle, and the bees go wherever wanderlust takes him. With a little ingenuity and research, bees can live, survive, and thrive in a wide assortment of places.

And wherever you do decide to place a hive, make sure it's out of the way of play areas, pets, and walkways. Placing hive entrances toward a solid structure, like a wall, will make the bees fly up and over the structure, effectively keeping them out of the way of people and pets.

MAKING ROOM IN THE CITY

Beekeeping in the country or on a farm has distinct advantages, such as space, sun, and water. But it wasn't that long ago in our history that most homes in towns also had at least one backyard beehive that provided honey during sugar shortages and **beeswax** for household candles and

quilting. Beekeeping in urban areas has become popular again, and one of the wonderful advantages is the proliferation of flowers. Bees often thrive better nestled in the warmer microclimate of a town, surrounded by beautifully landscaped flower beds or large community gardens where they can make abundant honey. This is in contrast to hives in the country, where the oversaturation of monocrops that are heavily maintained with chemicals can provide poor nutrition for bees.

If you plan to keep your bees in the city, there are a few things to consider. Make sure that there are plentiful flowers in your area through proximity to a park, lake, or natural area with wildflowers and flowering trees and bushes. Bees need fresh water, so if there are no natural sources close by, you will need to provide fresh water either through hive feeders or birdbath-type structures. If you plan on putting your bees on a rooftop or balcony, consider the weight of the hive on the structure itself, as well as your own ability to lift the weight of the honey down ladders or stairs. And consider how windy it gets and how low the temperature can drop on a cold rooftop or how scorching hot it may become in the summer without a source of shade.

PLANTING A BEE-FRIENDLY GARDEN

When we first moved to our country home, we enjoyed having the lawn mown on a regular basis and went to a great deal of trouble keeping it weed-free. After a trip abroad followed by a surgery, we fell behind on the mowing and weeding, resulting in a yard that began to grow in areas where it hadn't grown tall in years. We were amazed to see all the insects that came into our yard, and we willingly chose to stop mowing to provide an insect sanctuary. Providing a haven for pollinators, including honey-bees, is a wonderful gift both to humans and insects, as well as our ecology. Creating spaces for pollinators greatly benefits insects, providing them with a source of food, nesting areas, and mating grounds, and providing us with food, flowers, and entertainment. Even if you choose not

to become a beekeeper, keeping a bee-friendly garden will vastly benefit everyone.

GARDENING TIPS

Planting a garden for pollinators is easy to do and greatly rewarding. If you choose to plant a wildflower pollinator garden from seed, most seed companies have a wildflower seed mix buffet already prepared for you. All it requires is a toss on a sunny, spring day followed by a good dosing of water to start a fragrant garden. If you are choosing to begin with plants instead, ask at the nursery for some of the best choices for your area, making sure to consider your regional planting zone. Check flower plant labels for sun, water, and soil needs. Because bees forage all spring, summer, and fall, provide plants that will bloom throughout the seasons. Be cautious with any chemicals, especially pesticides on your plants, which can be taken back to a hive by a foraging bee, setting off a chain reaction that eventually kills the hive. Instead, kill weeds by hand and apply pesticides and other chemicals in spray form in the evening to allow them time to dry before bees begin their morning forages.

BEST BLOOMS FOR BEES

Honeybees are not choosy when it comes to flowers, enjoying ordinary clover, dandelion, and ragweed as much as they do the more exotic lavender and sages. Rather, bees pick flowers based not only on the size of their tongue (or proboscis), but also on the scent and color. In the spring, bees prefer flowering trees and plants like black locust and wisteria, which makes a very light-colored honey that has a light, floral taste. In the summer, bees love echinacea (coneflowers), Russian sage, and olive, as well as bee balm, zinnias, and cosmos. Now the honey begins to take on a more amber coloring with a more robust and full-flavored taste. In the fall, bee preferences go to sunflowers, goldenrod, yarrow, and sedums, which provide nectar that produces a very deep, rich, dark honey that tastes faintly of molasses.

PROBOSCIS

And don't forget, bees love your blooming vegetable gardens: green beans, peas, mints, chives, sages, herbs, tomatoes, peppers, watermelon, cucumbers, pumpkins, and many more. Allowing some of your garden plants, such as lettuces and coriander, to go to seed will thrill a bee. Honeybees primarily pollinate orchard fruit trees and nuts as well as berries. Bee-friendly trees for your yard include crabapple, dogwood, crepe myrtle, and flowering cherry.

You may wish to keep your bees to only a single floral source hoping to capture a special flavor in the anticipated honey, but monocrops do not provide an adequate diet for honeybees. Bees will fly to multiple sources of flowers when at all possible, resulting in different flower nectars in a single frame of honey, melding the various flavors into a rich, mellow, sweet floral taste.

SAFETY

One of the things we do that we really enjoy is teaching beekeeping classes. In the past 10 years, we have hosted and taught hundreds upon hundreds of beekeepers, both those new to beekeeping and those advanced in the craft. On our field days, we host another couple hundred people, and we host several school trips each year to our honey farm. In all those visits by those hundreds if not thousands of people, we have never experienced any kind of real bee-related emergency. Yet although negative incidents are rare, safety is a serious concern, and being prepared to deal with emergencies is important.

First, let's consider your yard. Keep children, friends, and your neighbors safe by erecting a warning sign so they don't accidentally come too close to a hive. For some, fencing in your hives could be a wise investment. Insist that visitors to your hives wear jackets, gloves, and beekeeping hats with veils before taking a peek. We tell visitors that it doesn't matter how much of a superman they are in their own hives; in our yard there are no shorts, no sandals, and no short-sleeve shirts.

If someone should get stung, be alert for any severe allergic reactions for at least one hour following the incident. Short-term, localized reactions to insect stings are common and can be dealt with quickly with ice and an analgesic, but anaphylactic shock is an emergency situation, and 911 should be called immediately. Always carry a cell phone with you in the bee yard.

Keep yourself safe in your yard by always wearing the proper beekeeping protection as well as using a smoker. Practice proper lifting techniques in lifting supers or moving hives. If necessary, get a friend to help and invest in a beekeeping lift bar for those heavy boxes. And remember your smoker gets hot, so always set it in a safe place when it is not in use.

PETS AND CHILDREN

You may have concerns about starting beekeeping because you have small children or grandchildren, along with various pets. We have been beekeeping for about 25 years, and during those years have raised six children, doted on over 11 grandchildren, raised around 50 chickens a year, and have had too many dogs and cats to count.

We have not noticed any issues with the animals around the honeybees. In fact, it is a common occurrence here to see honeybees, dogs, and chickens all communally drinking from the same water dish. Friends of ours keep their hives in their horse paddock, with the horses. Another keeps his hives along the fence that contains his goats. However, animals do get stung just like humans, so for your pets' safety and yours, keep animals away when you are out in the beehives.

Teach small children to stay back from the hives, never playing, climbing, or pounding on the boxes, and never sticking their fingers into the hive entrances. Always insist that children wear shoes outside, and if you choose to allow young children to be near the hive with you, invest in child-size protective gear and gloves.

PROTECTING AGAINST NATURAL PREDATORS

Hives are not only agitated by small humans, but by animals as well. Insectivores like skunks will bother hives at night by pawing into the entrances. Placing hives high on stands or using tack strips at the entrance will deter this animal. Keep a large rock or brick on hive lids to keep the crafty raccoon from lifting it off. Bears can also be very detrimental to hives, destroying not only the bees, honey, and comb, but usually the wooden equipment as well, representing a financial and often emotional loss. Erecting a livestock fence, much like the ones used to keep in chickens or sheep, around your hives should deter bears. Most other animals seem indifferent to bee hives, although if you are in a high-traffic area for animal crossings, you may see some tumbled lids and covers now and again.

EQUIPMENT AND TOOLS

Aside from the beehive and the honeybees, a beekeeper will need to invest in some primary beekeeping equipment. Beekeeping is like any hobby in which you will find that there are hundreds and hundreds of gadgets and gizmos to purchase. Yet a beekeeper needs just a few simple items to get started, such as a hive tool, smoker, and protective gear like a hat, veil, jacket, and gloves. Invest in a sturdy bucket or toolbox for your beekeeping tools, making sure it can be easily transported to and from your hives. For more information on where to purchase these items, check out our Resources section.

SMOKER SUPPLIES

A beekeeper uses a smoker while inspecting a hive. The microscopic smoke particles from a beekeeper's smoker disrupt the pheromones that bees use to communicate, allowing the beekeeper to open and inspect a hive. When purchasing a smoker, look for one made with heavy, durable stainless steel. It should also contain a metal heat protection cage or heat shield and have quality bellows made of vinyl or leather.

Smoker fuel can be purchased commercially and consists generally of paper and cotton products or compressed wood pellets. There is also a wide array of natural items you could collect to use as smoker fuel as well, such as bark, mulch, pine needles, and dried grasses. Twine, burlap, and cardboard formed into tight rolls can be used as well. When working with any smoking equipment, be safe and have a fire extinguisher on hand in case the flames get out of hand.

GEAR

Besides a smoker, the most important item a beekeeper should invest in is a beekeeping jacket with a veil, a full-length beekeeping suit, or a hat with a veil. No protective gear is sting-proof, but having a jacket or suit with thick, strong material can help protect you from unwanted stings. Beekeeping outfits are traditionally white, but suits and jackets can now be found in a wide assortment of colors and designs.

Gloves for beekeeping are a must for a beginner and should reach to your elbows and have elasticized armbands. Gloves are made from leather or canvas. Some beekeepers switch to nitrile or chemical gloves when they are dealing with a lot of supers and sticky honey, propolis, and wax.

If you choose to forgo the traditional beekeeping wear and go with only a hat with veil, remember that you can protect yourself additionally by wearing long pants and closed-toe shoes, taking care that sleeves and pant leg openings are closed to wandering marauders. Our eldest son uses masking tape around his arms and legs to keep out bees, and although it's a comical sight, it is effective.

FIRST AID

We always keep a first aid kit with us when we take students out into the yard. In addition to some of the basic items in a kit, such as gauze, bandages, and antibiotic ointment, our kit includes sting wipes, which are pads presoaked with benzocaine, a topical anesthetic used to relieve the pain of stings. There are many old wives' tales about relieving bee stings with meat tenderizers, toothpaste, or even baking soda, but I've never found anything that worked as well as ice, so we always carry a cold pack with us. Finish by drying off the area and applying a hydrocortisone cream or gel.

OTHER SAFETY EQUIPMENT

Hive tool. After the smoker and protective gear, the single most important tool a beekeeper needs is a hive tool. A hive tool is a type of pry bar that the beekeeper uses to pull apart the hive components (which are stuck together by the bees with a sticky substance called propolis). Hive tools come in various colors and lengths, from a few inches long to 10 inches. Buy a bright, neon-colored tool if you're prone to dropping them in tall grass!

Feeders. Insects don't need to be fed like animals do, but all new packages of bees started in beehives in the spring generally need to be fed with sugar water and protein because floral sources are still scarce. Feeders greatly help with the new production of wax, which will be the biggest job a new hive has to begin. Hives should be fed again in the fall as floral sources begin to wane. A beekeeper will want to have available feeders for each hive. There are many different kinds of feeders, including both inside- and outside-of-the-hive feeders. Feeders used outside the hive (Boardman feeders) are inexpensive and easy to maintain but can induce robbing from neighboring hives. Inside-hive feeders are more cumbersome and a tad trickier, but hold more sugar water so there's less maintenance.

Excluder. You may choose to use a queen excluder on your hive. An excluder is a plastic or wire mesh that is placed between a deep super and the honey super to hold your queen below in her brood boxes, which keeps her from laying eggs in your honey.

See Resources for more information on where to order equipment.

REGISTER YOUR HIVE

Most beekeepers may wonder if they need to register beehives within their community or with their state. It may be legal to keep bees within your state, but backyard beekeepers also have to check their town or municipality's laws and regulations concerning beekeeping. Some towns have nothing on the books about beekeeping, but others have very strict regulations on where you can keep beehives within city limits, how many you can have, and how far the hives must be from a neighbor. Some communities also now require beekeeping certification, showing that you have taken training designed to teach the process and requirements of beekeeping.

States require the registration of beehives in order to effectively track and control disease outbreaks, communicate with pesticide sprayers, and alert aerial chemical applicators. Every state has different registration requirements; ranging from simple paperwork to detailed information and maps of apiaries. Check on your state laws by calling or checking with your State Department of Agriculture or Department of Natural Resources. Most registration forms are simple to fill out and may require just the basics like your name, address, and latitude and longitude coordinates. Easily find your coordinates by selecting and holding down your area on the Google Maps app.

Many states do not charge a registration fee, but some may charge a modest fee for hive registrations. You may need to register or update your information annually. Some states retain the right to levy fines or fees if you choose not to comply with state laws in the area.

COMMON STATE LAWS

Registration required	Most states require registration except: AL, CO, FL, HI, IN, KY, LA, ME, MS, MO, NC, SC, OH, VT, WI
Apiary inspection required	Most states require inspection except: AL, CO, CT, HI, MO, SC, WA, WI
Apiary inspector has right of entry	Required in all states except: AL, HI, MO, SC
Inspection certificate required	Most states except: AL, CO, HI, ID, IN, IA, MD, MI, MO, MT, NV, NH, NY, RI, VA, VT, WA, WI
Hives must have removable frames	Required in all states except: AL, CT, HI, LA, ME, MS, MO

Note: Check with your own local and state governments for updated laws and regulations.

PART 2

Caring for Your Bees

Honeybees deserve our best management practices. Once we understand the significant value bees play in our ecosystem, we soon appreciate caring for our bees as the amazing creatures they are. Their natural habitat is a hollow tree. However, due to the increase of pests and diseases, a domesticated colony in a man-made hive has a far better chance at survival than a feral hive. A managed colony can be inspected regularly, and action can be taken to improve the overall health at the first sign of a problem.

As you begin your backyard beekeeping endeavor, you will be given many opportunities to give your bees the very best care, from the assembly of your hive equipment to transporting your bees home, to properly installing them into the hive. You will learn to feed your bees at the right time and, of course, caring for your queen is paramount. You will grow to love your hive inspections, giving you the opportunity to "groom" your bees and evaluate their needs.

SETTING UP YOUR FIRST HIVE

O ur very first hive was a cutout from a fallen tree. That was rough. Our next hives were set up the traditional way by purchasing a package of bees and installing them into a hive. We started long before anyone sold assembled hives, so we purchased our first hive by mail order. Weeks later it arrived in hundreds of pieces. There were staples, nails, glue, wires, and precut boards that could only fit one certain way. We had no idea what we were doing, and we were so excited that we placed one board on the wrong way. Unfortunately, it meant the handle was on the inside of the hive. We followed some mimeographed printouts on how to install a package of bees. There was no Internet or YouTube. At the time, we felt like we knew what we were doing, but we had to have made a ton of mistakes. Today, new beekeepers have many more opportunities to start with greater understanding.

HIVE ASSEMBLY

It is best to purchase your hive long before you pick up your bees. This will give you plenty of time to assemble your hive, paint it, and place it where you want it to be in your yard. Today, most beginners purchase hives that are already completely assembled and painted. Occasionally, someone might enjoy woodworking and have all the tools, but for the most part, starting with an assembled hive has so many advantages. An assembled hive means that the individual parts are constructed together. However, you still must place the parts in the proper order. If you purchase an unassembled hive, it will take you one to two days to assemble all the parts and frames. If you purchase an assembled and painted hive, it will take you 30 minutes to place it in your yard. Once your hive is in place, you can easily bring your bees home and install them in your new hive.

LANGSTROTH HIVE

 If you purchase an unassembled Langstroth hive, it means that you are very good at building and woodworking, so simply follow the instructions. Since most Langstroth hives arrive assembled, you simply need to stack your pieces in the proper order on your hive stand.

1. Assemble the bottom board as the first piece for the hive of your choice. The bottom board must be placed with the largest edge rail-side facing up.

2. Next, assemble your deep hive box and the frames and foundation. Place the frames in the deep hive box and place it on top of the bottom board.

3. Repeat the deep hive box for your second hive body and place it on the first one.

4. Next, assemble your honey super and the frames and foundation. Place the frames in the honey super and place it on top of the top deep hive body.

5. Assemble your inner cover and place it above your honey super.

6. Lastly, place the top cover above your inner cover.

You will not use all the boxes when you first install your bees as this would give your bees too much room. You want to encourage them to start adding wax to five or six frames in one box before you allow them to advance into the next box.

TOP BAR

If you purchase an unassembled top bar hive, it is likely that you are handy with woodworking and you will be building your top bar hive based on your particular size and available material specific to your needs. You may find it more convenient to purchase your top bar hive mostly assembled. You will likely need to attach the legs to your top bar hive. Most legs bolt onto the main body. Once your legs are mounted, you can position your top bars. Then your slanted roof is placed on the body of the hive.

1. Assemble legs with bolts onto the hive body.

2. Position top bars into the groove inside on the top of the hive body. Many top bars are angled and come to a point. Others have a small slat of wood that slides into the bottom of the top bar. It is best to always coat the slat or tip of the angle with beeswax to give the bees a good start to attaching and building their comb to your top bars.

3. You will need a top bar divider. This is a top bar that has an entire piece of wood shaped to the same angle as the hive body. A divider is used to reduce the size of the cavity in which the bees can work. When first installing your bees, a divider is needed to keep the bees within an area containing only 5 to 10 frames. This will ensure the bees will quickly begin building comb onto the top bars. As these frames are started, the divider bar can be moved to now include 5 or 10 more frames.

4. The top bar roof should always be placed on the hive to shield it from the elements. We prefer to use fasteners or tie-down straps to hold our top down on top bar hives because the angled roof will not accommodate a cinder block or weight.

A nuc is a nucleus of a larger hive, and if you install one, it means you receive four or five frames from a healthy colony with the queen and move them into your new hive. Starting with an **established colony** means the hive is completely operational and is usually operating at maximum population. Installing a nuc into a top bar hive is nearly impossible due to the nonproprietary size of top bar hives. If you did find someone willing to sell you five frames of comb and bees from their top bar hive, it is likely their size and shape of comb will not fit your top bar hive. Even the frame length is usually unique to the builder. It's hard to find or purchase a complete top bar hive with bees included and working as a full colony. Both the nuc and the complete colony would also require that the bees be fully inspected prior to purchase since moving comb can transfer diseases.

WARRÉ HIVE

If you purchase an unassembled Warré hive, follow the manufacturer's recommended assembly instructions. Many Warré hives come mostly assembled. Once your pieces are assembled, it is a matter of placing the pieces in the correct order. The Langstroth and top bar hive all work best when you begin with minimal space. A Warré is no exception. The bees are started in one deep hive box of a Warré hive. Once the bees have nearly drawn the comb out on five or six frames, the next box is added below the first box. In true Warré fashion, the idea is that bees prefer to build downward as in their natural habitat. Boxes are added below as each box above has five or six drawn-out combs. This continues until the hive reaches the space needed to accommodate a large colony, usually three to six boxes. If the colony performs as they do in nature, eventually the top boxes will become filled with honey and the beekeeper is free to harvest honey from the top super once the comb is capped over with wax, sealing in the precious honey.

1. Place the bottom board on a stand.
2. Add boxes below the current box as needed.
3. Place the quilt box on the top of the hive.
4. Place the top cover on top of the quilt box.

BRING THE BEES HOME

If possible, it is best to pick up your bees from a package bee provider. Shipping can be stressful on bees. Every year we have a day when we bring in our packages for customers to pick up. The air is filled with optimism and elation. In all the excitement, many new beekeepers make mistakes when picking up their bees, so let us give you some pointers.

Always take a hat and veil. Even though the bees are in cages, there are always bees on the outside trying to get in. We call these "hitchhikers." When we are packaging bees in the bee yards, many bees are flying around and some land on the packages, trying to get in with the other bees. So, expect bees to be flying around when you pick up your bees. Never place your bees in a box or seal them off. It's natural when traveling with bees to want to keep them from getting out in your car. But cutting off their air will kill the bees. If some get loose in the car, they will fly to the windows to get out. Bring sugar water mixed with one part water and one part sugar to spray on the outside of the cage prior to your journey home. This will settle the bees down during transportation and help them remain well-nourished while being caged. Always carefully inspect your bees before you pay. Make sure there are very few dead bees on the

bottom of the cage. Also, ask to be shown your queen in her cage to ensure she is alive and moving around.

INSTALLING YOUR BEES

Installing your package of bees may seem overwhelming, but it is a beautiful experience. Bees are not very defensive in a package. Remove five frames out of your deep box. Open your package and remove the queen cage and set it aside. Spray the outside of the package of bees with 1:1 sugar water, then lift the package several inches from a surface and slap it down to cause the bees to be knocked to the bottom of the cage. This will allow the bees to be more easily shaken out. Now, start shaking the package of bees into the deep box into the space you created by removing the five center deep frames. Continue to knock the package on the ground or table to jar the bees toward the bottom of the cage again and shake them into the hive. You will need to repeat this several times. Next, place the five frames back into the hive, being careful not to smash any bees that may be piled up on the bottom board. Finally, remove the cap or cork covering the candy on the queen cage and place her cage in the center of the hive box between the two center frames. Now, place your inner cover and top cover on. Be sure to place either an entrance feeder or a top feeder on the hive, as this new colony will take several days before they begin bringing in resources.

INSTALLING THE QUEEN

When installing a package of bees, the queen will come in her own cage, sometimes with **attendant bees** and sometimes not. Look closely, and you will notice that part of the cage is plugged with a white candy. There is usually a plastic cap or a cork protecting the candy from the outside. Remove the cork or cap. Once you shake your bees into your hive, it is best to suspend the queen cage between the top of the frames in the middle of the hive box. To do this, you can use a small, stiff wire or a paper clip and bend it to go through the cage and hook over the top of a frame. The cage should hang just below the top bar of the middle frames. We prefer to place the hole with the candy facing down so the queen can walk downward out onto the comb rather than up on the top of frames. If the candy plug seems very hard or very thick, you can remove half of the candy to speed up her release. Do not remove it completely. Do not remove the wrong cork or cap that will release her instantly, only the candy side plug. If she is released too quickly, the bees may not accept her. They need time to become acquainted with her pheromones.

BEES IN THE YARD

Being surrounded by neighbors means that the less conspicuous your bees are, the better. It is impossible to keep your bees on your property because bees fly miles away to gather their choice of optimal nectar. However, there are several techniques that you can use to reduce the threat of your bees alarming your neighbors. Most of the year, your bees will go about their activities unnoticed by your neighbors. However, the fall nectar dearth sends bees searching for sweets. To keep your bees from your neighbor's trash or picnic, provide top feeders on your hives. You can also place pans of sugar water in your yard so that bees will prefer to stay on your side of the fence.

You can paint your hives colors that blend into your yard or garden so as not to draw attention. Probably the most positive impact you can have on your neighbors is to prevent your bees from overcrowding and swarming onto their trees or house. It can be difficult to prevent swarming altogether, but always giving your bees room to grow, splitting hives in the spring, and keeping a young queen in the hive can help greatly reduce spring swarming.

FEED THE BEES

Most everything bees eat they eat through a straw: their proboscis, which is essentially their tongue. They can shape it into a straw to suck nectar from flowers or fresh water from puddles. Bees consume the same foods we eat: carbohydrates, proteins, vitamins, minerals, and water. Many beekeepers mistakenly assume all bees need is sugar. A sugar-only diet is not healthy for bees, as they need protein as well.

When you first start your new colony, feed your bees sugar water and include protein powder mixed into the sugar water. When bees consume flower nectar, they are also consuming protein through the pollen that makes its way into the nectar. There is no need to feed bees when they are established and foraging for nectar and filling honey supers. However, as soon as the summer nectar flow ends, begin feeding your bees 1:1 sugar water with protein again. In the winter, feed them a hard candy board containing sugar and protein. Except in the winter when bees cluster, provide water around your yard. A birdbath with wooden sticks or rocks will provide the water your hive needs to keep cool in the summer. A quart of 1:1 sugar water requires 1 teaspoon of bee protein powder, which can be purchased at most bee stores. It is also helpful to add 1 teaspoon of Honey-Bee-Healthy to stimulate the bees to consume your feed. We have found Honey-Bee-Healthy very helpful. It is a feeding stimulant liquid that is added to sugar water for bees. It's made up mostly of spearmint water and lemongrass oil.

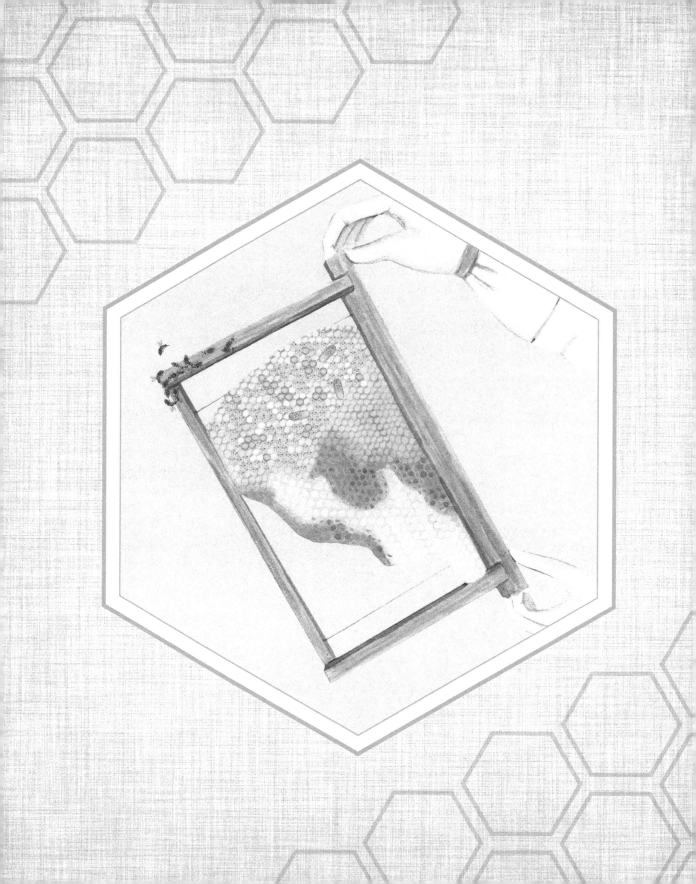

INSPECTING YOUR HIVE

One of the more enjoyable aspects of beekeeping is the thrill of inspecting the hive, which offers a chance to watch your bees go about their many activities. When we were new beekeepers, we had no idea what we were doing or what we should be looking for. Your hive inspection should be a time to enjoy your bees up close and personal while also observing their overall health. Be sure to gather up everything you will need before you open your hive. Always use a smoker to calm your bees. You will be looking at very tiny eggs, so choose the proper visual assistance to help you see clearly. We use reading glasses to help us see up close. You may be nervous your first time, unsure of what to expect, so wear the proper amount of protective gear to help build your confidence. Stay calm and enjoy the experience. How many people have the privilege of watching honeybees work up close? Work your inspection from the back or side. If you perform your inspection from the front, you will hinder the flight of foragers coming and going.

TIMING

Develop a regular routine for inspecting your hive every two weeks. You do not have to perform a meticulous and thorough inspection that takes hours. Rather, you're specifically observing the productivity of your queen. You must verify that you see an abundance of eggs. When you see eggs, you know the queen was there to lay those eggs within the last day or two. Performing an inspection every two weeks confirms that your queen is healthy and has a great brood pattern. If you lose your queen and the colony fails to raise a new one, then after three weeks the hive begins spiraling into a tailspin, and it becomes more difficult to save. Inspecting every two weeks allows you to catch any queen issues in time to place a new queen in the hive before the hive begins to fail. Your main goal is to verify eggs and capped-over pupae in the colony.

WHEN TO INSPECT

After you install your package, you will be excited to inspect the hive to see if the queen has been released from her cage. It is important to wait at least three days to allow the queen to exit her cage and become acquainted with the colony. If she is not released from the cage, you can either place the cage back in the hive or open the cage and allow her to walk out onto the frames. The queen's pheromone is an important factor in helping solidify the colony as a single superorganism. Be patient. The bees need time to eat through the candy in the queen cage to release her and to start building some comb on the new foundation.

WHEN NOT TO CHECK

Our busy lives may interfere with the timing of inspecting our hive. For example, we may only have time in the morning before we leave for work or when we get home after work. This may not work. Inspecting bees too early in the day or too late in the day is not the best time. Bees are not foraging at these times, resulting in more bees being home in their hive.

Avoid opening a hive for inspection when it is dark, raining, or during high winds. Never inspect your hive unless the temperature is above 60 degrees Fahrenheit.

HOW OFTEN TO CHECK

Looking in on your bees is one of the most fun and enjoyable aspects of beekeeping. Many new beekeepers worry that they are not inspecting enough or that they may be inspecting too often. What is the best approach in planning your hive inspections? Every two weeks is the happy medium. With a colony, we must be able to detect any issues early. Inspecting our hives every two weeks will give us the opportunity to address a problem before it gets out of hand. Frequent inspections will allow us to monitor how well our queen is laying eggs. Since the queen lays over 1,000 eggs a day, inspections that are too far apart could cause the colony to be very low in population from a failing queen. Inspecting every two weeks gives us an edge on making corrections by replacing the queen should she start to fail. It also allows us to know how well the foundation is being drawn out with wax or if we need to create more space by adding boxes to the hive. An inspection every two weeks also helps us keep an eye on any pests or diseases before they spread out of control.

CHECKING TOO MUCH

Can you inspect the hive too often? Maybe, especially if your bees are extremely agitated by frequent inspections. Certainly, you can learn the art of inspecting a colony by keeping the bees as calm as possible. Until you develop this skill, limit your inspections to every two weeks, because a new beekeeper may squish and kill bees with clumsy handling of the frames. You do not want to accidentally kill your queen. Each inspection does break the propolis seal of the top cover. However, on warm days, it is quickly resealed by the weight of the top cover. It will not reseal on cooler days.

OBSERVING

Much can be observed when walking up to the hive, even before opening the top. Look for anything unusual around the hive, such as an excessive number of dead bees in front of the hive. A small amount is normal. Also look for any unusual signs of animal tracks, such as those from skunks. Skunks will leave scratch marks on the ground and on the front of the hive. Look for bugs, such as yellow jackets, spiders, or moths. Listen to the sound of the bees. Can you hear a normal low hum? Are the bees moving normally in and out of the front entrance? Good, that's normal. If they are being robbed by another hive, there will be bees all around the hive. Are there any foul odors around the hive?

USING A SMOKER

The smoker is the beekeeper's best friend as it calms bees prior to and during an inspection. Folklore says that smoke causes bees to believe their hive is on fire, causing them to gorge themselves on honey in case they need to leave their burning hive, and when their stomachs are full of honey they cannot sting. However, science shows that bees communicate through pheromones, and smoke masks the pheromones. When bees lose their ability to communicate, they are calm. Not only does using a smoker reduce stings, but fewer bees will die from being defensive. Honeybees die when they lose their stinger and second gut. The more you can minimize any defensive behavior, the more bees you keep from dying. A calmer hive inspection can also prevent defensive bees from picking on your neighbors.

Always start the smoker with flammable material from the bottom of the smoker first, like newspaper. Slowly add more smoker fuel with greater substance, such as burlap, pine needles, hemp rope, unprocessed cotton, or even old blue jeans. Avoid fuel that may contain contaminants such as from cardboard, polyester clothing, or toxic chemicals. Use natural smoker fuel. At first, you will need to start a strong fire in the smoker's canister. Try not to smother the fire by adding too much additional fuel. Once the smoker is burning well, add a bit more fuel and close the lid. Once you close the lid, you will need to puff your smoker frequently to continue to feed air to your enclosed fuel. A lack of puffing from the bellows will allow the fire to go out, and you will not have smoke when you need it most. Extinguish your smoker carefully when finished to avoid starting a grass fire. Some smokers do not have a grille around the canister and can burn your hands, so be careful to only hold a smoker by the top of the bellows.

INSPECTION PROCEDURE

As you prepare for your hive inspection, you will need to be well prepared so you can be as thorough as possible as quickly as possible. You do not have to rush through an inspection, but you cannot afford to keep the hive open more than thirty minutes. The longer the hive is open, the more likely the bees will lose their patience. Prepare how you will perform your inspection and what you will be looking for and doing once you open the hive. Have a plan and work deliberately. Lay a checklist on the ground near the hive to remind you of key inspection points. Harvesting honey will be covered in chapter 9.

GETTING READY TO INSPECT

It is important to prepare properly for your inspection. Locate all your equipment in advance, such as your protective gear, hive tool, smoker and smoker fuel, lighter, glasses or magnifying glasses, bottle of water, and whatever else you may need. It's a frustrating feeling when you think you have everything only to find once you open the hive that you forgot your glasses or your hive tool. Take a hammer and a pair of pliers for that nail that is sticking up or frame that may have pulled apart. You can take a drink of water without removing your hat by drinking through the veil of your mask.

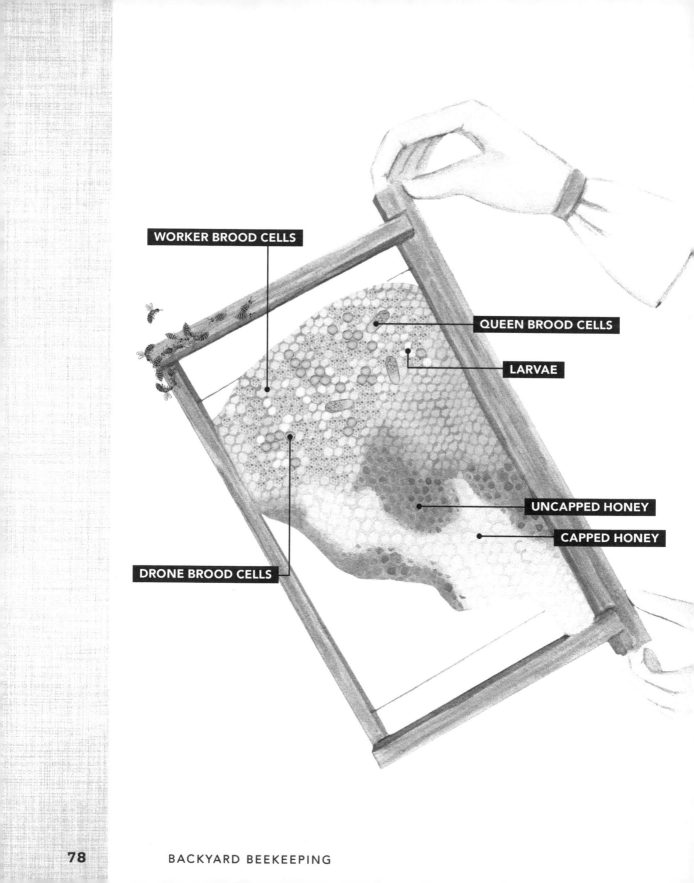

WORKER BROOD CELLS

QUEEN BROOD CELLS

LARVAE

UNCAPPED HONEY

CAPPED HONEY

DRONE BROOD CELLS

OPENING THE HIVE

Upon approaching the hive, have your smoker burning well and give several puffs of smoke near the entrance to calm the guard bees. Now, walk to the side or the back of the hive and use your hive tool to gently lift the top cover about two inches and give several puffs of smoke under the top cover. Set the top cover back down for a few seconds to allow the smoke to drift down on the bees. Next, remove the top cover and repeat the same process with the inner cover, giving a few puffs under it, then removing it, always moving in slow motion. Set the top cover and inner cover against the hive or in your yard. Give a few puffs of smoke over the bees that are now exposed on the top of the frames. If you need to remove your super, insert your hive tool under the super, lift about an inch or two, and give a few puffs of smoke, then pick it up and set it onto your upside-down top cover on the ground next to the hive. To lift frames, choose the first one closest to the hive wall and gently and slowly lift it up, setting it outside the hive to allow space to slide other frames.

CHECKLIST FOR INSPECTING

☐ **Population Growth.** Look for the overall population growth of the colony.

☐ **Drawn Comb.** Observe the condition of the frames such as if comb is being added to new frames.

☐ **Content of Frames.** Take note of what each frame contains, such as brood, bees, honey, and pollen.

☐ **Is the Queen Laying?** See if you can spot your queen, verifying that your queen is in good shape. This will ensure that your queen is laying well.

☐ **Overall Observation.** Make a quick glance for anything unusual such as a declining population, lack of eggs, or queen cells. Queen cells are peanut-sized cells hanging vertical to the comb housing developing queens. This may indicate the queen is being replaced or the hive is preparing to swarm.

INSPECTING FRAMES

When you lift a frame up, be sure you have a tight grip on each of the side ears of the frame. Do not drop a frame of bees. Begin observing the frame. See how much wax has been added to the foundation and look to see if the bees have placed any nectar or pollen in the cells. Can you see eggs or larvae in the brood area? If you see eggs, there is a good chance the queen is nearby. Start scanning the frame for the queen. She is difficult to distinguish from the worker bees at first, but this will get easier the more you see her. She is longer, and if she is marked, it will be much easier.

As a new beekeeper, it will be challenging to distinguish between capped-over brood, pupae, and capped-over honey. To a new beekeeper, capped-over brood and capped-over honey look the same on a frame. However, capped-over honey has a more wet, waxy look. Capped-over brood has a drier, more velvety-looking cap. If you are not sure, you can use a toothpick to peel back the capping. You will either see two purple eyes looking up at you, or honey. Now you know the difference.

Pests and diseases are more difficult to diagnose visually in the hive, but there are some symptoms to look for, such as those that can indicate American foulbrood, a highly contagious brood disease. The capped-over brood is sunken and perforated, and the comb has a disgusting smell. Also keep an eye out for wax moths and small hive beetles.

REPLACING FRAMES

After several years of keeping bees, brood frames will become aged and may need to be replaced. In the brood area, a pupa spins a cocoon in each cell for each generation. After many generations, the cells become lined with cocoons and are smaller in size. Some diseases can grow between the layers of cocoon. Changing out the frames every five or six years can help prevent the spread of some of these diseases. Many beekeepers will change out their foundation when the comb becomes very dark, almost black with age and use.

INSPECTIONS BY HIVE TYPE

Hive inspection fundamentals are similar for different hive types, but there are some differences that are noteworthy. These differences between hive types are important, because they may help you in performing your hive inspection safely. Differences include how to open a hive, how to handle frames, and how to keep the bees calm. Once you have the inspection fundamentals down, it will be easy to adapt these to different types of hives.

LANGSTROTH

The traditional Langstroth hive inspection involves smoking the entrance, under the top cover, and under the inner cover before removing components. Allow a minute or two for the smoke to calm the bees. Remove the top and inner cover slowly. Smoke the top of the frames once the inner cover is removed. A Langstroth will usually include a wooden inner cover under the top cover. Also, Langstroth hives have flat top covers, which make them handy to place upside down next to the hive during inspections to set boxes on. A Langstroth also accommodates a queen excluder

between the brood nest area and the honey super, keeping the queen from laying eggs up in the honey super. Bottom boards can be made of solid wood or contain a screen bottom for ventilation and mite control.

Since the Langstroth frames are reinforced with wood on all four sides, handling a frame during an inspection is very convenient. Frames are held by the outer edges and can be rotated or flipped to inspect both sides of the frame.

TOP BAR

The top bar hive inspection is different than that of a Langstroth hive in many key areas. The most obvious is the handling of frames during an inspection. Since a top bar frame consists only of a top wooden bar, the comb is only suspended from the top bar. If the frame is lifted and held sideways, as with a Langstroth hive, the top bar comb will break off. The frame must always be held vertically. This means that after inspecting one side of the frame, it must be rotated 45 degrees, then rotated so the comb is always vertical. Now you can view the other side. If you hold it perpendicular to the ground, the comb can break loose.

WARRÉ HIVE

A Warré hive still requires the same timely inspections as other types of hives. However, many Warré beekeepers feel it is best to inspect as little as possible so as not to disturb the colony. Minimizing inspections seems respectful to the colony, but a colony must be inspected regularly; regular inspections are far less harmful than pests and diseases that may go unnoticed due to fewer inspections. Depending on the type of frames in a Warré hive, they may be similar to the Langstroth frames, consisting of a foundation surrounded by four pieces of wood, or top bar frames, consisting of only a top bar holding the comb. When inspecting a Warré hive, the top will be removed first, then the quilt box, and finally the cloth inner cover. The actual observation of the bees and their health is similar for all types of hives.

GETTING STUNG

Honeybees are beautiful to observe as they go about their majestic activities. However, because of the precious honey in their hive, they have a stinger to keep intruders from stealing their liquid gold. Occasionally, while inspecting a hive, beekeepers will encounter a sting or two. If you are stung, remain calm. Slowly place the frame back into the hive. Do not panic or scream or run. This may excite other bees. Smoke the hive with several puffs of smoke, back a few feet away from the hive, and use your hive tool to scrape out the stinger. Then, smoke the area where you were stung to reduce the alarm odor that may have been left by the stinger. Have a supply of ointment or salve to place on the sting area, which can greatly reduce the pain. The sooner you remove the stinger, the less venom will enter the sting area. Many beekeepers will use honey, plantain leaves, and other home remedies to calm the sting area.

ANATOMY OF A STINGER

The stinger on the honeybee is about 1.5 mm long; very small. A microscopic look at the stinger reveals that it consists of two serrated blades that move opposite each other, which allows the stinger to penetrate. Venom is then pumped from the venom sac attached above the stinger through the center of the stinger. Because the venom sac is attached to the intestines and gut of the bee, and because the barbed stinger stays in the victim, when the bee tries to fly away, she pulls out her intestinal tract. Within an hour, she will die. The venom sac is visible as a yellow blob on top of the stinger. This venom sac releases an alarm pheromone, helping other bees to locate the intruder. Removing the stinger quickly will greatly reduce the amount of venom that is pumped through the stinger, thus reducing the local reaction.

KEEP A BEEKEEPING JOURNAL

Documenting your experience as a beekeeper is of tremendous value in evaluating your methods and practices. After each hive inspection, keep a journal of everything you observed. Record what you did and what you saw. You must keep careful records about your queen's progress. Did you see plenty of eggs and capped brood? Are the frames being filled with comb? Are the bees bringing and storing pollen and nectar in the comb? When you perform your mite tests, keep careful records on the mite count and any effort you made to reduce mites.

Also, keep a checklist of items you should observe. Have the list handy during your inspection. Otherwise, you may become so focused on finding the queen that you fail to observe her laying pattern. Also, after the inspection, write down any proactive future actions you must take during your next inspection.

Many people keep journals in notebooks. There are also many mobile apps to assist you in keeping records of your hive inspections. Hive Tracks is software that many beekeepers enjoy that allows you to collect and share data from your hives.

YOUR COLONY BY THE SEASON

Your new colony will continue to develop, expand, and grow throughout the seasons and throughout the next few years. During the first year, the colony will start off very small and seem very easy to manage and observe. However, the colony will begin to expand quickly as the queen lays over 1,000 eggs a day. The first year is a learning period of observing and working to grasp keeping bees. The second year will bring a greater sense of understanding. The hive will be able to expand rapidly the second year as all the frames are drawn out from last year, starting off the second year far ahead of where they were a year ago.

FIRST FEW MONTHS

During your first few months as a new beekeeper, you will be filled with excitement and enthusiasm. Your bees will fascinate and entertain you. They will also leave you questioning what's going on, why are they doing this or that. But as you continue to inspect your hive and become familiar with their activity and behavior, your confidence will build. It is best to inspect your hive every two weeks during the first few months. This will not only help you keep close tabs on your new hive's progression, but will also allow you to learn and become more aware of the wonderful hobby of beekeeping. During the first few months, the hive will be small in population, between 10,000 and 20,000, so you will enjoy the ease of working a smaller hive. You will also enjoy watching your bees expand in population and into more boxes.

SPRING, YEAR ONE

What should you be seeing during your first spring? Your chances of finding the queen during your first spring will be greater because the population of worker bees is much lower. You will enjoy finding the queen and observing eggs in the cells, confirming that all is well, and that your bees are well on their way to becoming a large and healthy colony. At first, the bees will take several days to become familiar with where they live. They will take many orientation flights. After several days, foragers will fly miles away and return to their exact hive without hesitation. Be patient with the honey harvest on your first year. The bees are working hard to use incoming nectar to make wax on all their new frames. The second year is when you will see more honey to harvest.

SPRING, YEAR TWO

During your second season, spring will be exciting for you. Bees will break from their winter cluster and begin cleaning up the hive from winter. You may see many dead bees from winter being removed from the hive. Occasionally, we will remove the bees from the entrance, allowing bees to fly in and out on warmer winter days. You will experience a strong colony coming out of winter. Bees will be flying out in a frenzy to gather pollen and nectar from the early spring fruit trees and gardens. Since your colony is now in its second year, it is far more advanced than it was last spring. With all the frames drawn, there is no limit on brood production, pollen and nectar gathering, or honey production.

Every healthy colony will instinctively reproduce another hive in the spring. This is called swarming, when half of the colony leaves the hive with the old queen to form a new home of their own far away. The old hive left behind will raise a new queen and carry on. This is how bees reproduce in the spring, making more colonies. Beekeepers can be prepared with extra empty hive equipment and capture these swarms and increase the hives in their **apiary**.

Spring can be damp with spring showers, so be sure to secure the top with a large rock or weight as tops are often blown off during spring thunderstorms, allowing rain to soak the bees.

SUMMER

During the heart of summer, the bees have reached their maximum population of around 40,000 and are fully engaged in foraging on clover, gardens, crops, and flowers in pastures. In climates of extreme heat, bees must keep the internal temperature of their colony around 92 degrees Fahrenheit. On extremely hot and humid days, the colony will shift from nectar foraging to gathering water. Foragers bring water into the hive and place droplets of water on the brood frame, then fan their wings to create an evaporative cooling effect, which keeps the developing brood from overheating. We will place a four-by-four piece of plywood on top of our hives to provide shade on extremely hot days. It is also helpful to remove the bottom entrance reducer and to place a small stick or coin under the top cover, raising it slightly for improved ventilation. Hives benefit from being placed in sunny areas, but afternoon shade can help keep the hive from overheating. This can be accomplished by placing a hive in a location that has sunlight in the morning and early afternoon but is in the shade in mid to late afternoon.

BEARDING

Bearding is when a large number of bees crawl outside their hive and gather on the entrance or in front of the hive to alleviate some internal heat. With the use of a screened bottom board, the bees beard less because the added ventilation helps the bees control the hive temperature. Bearding occurs mostly on hot and humid evenings. Often the colony spreads out on the ground in front of the hive. Keep a water source near your hives, such as water in birdbaths or in large pans with some floaties to keep bees from drowning in the water. Rocks placed in pans of water will give the bees a place to land and retrieve water for their hive.

FALL

As summer winds down, there will be a dearth, a time when nectar sources are drying up as fall approaches. For us, bees can be foraging heavily and then in one day it is as if someone turned off a switch and the bees are done foraging. This is a great time to harvest honey from the hive. You will likely see your honey supers full of capped honey. This is worth noting, and when it happens, the beekeeper should immediately begin feeding their bees. Never use an entrance feeder in the fall, as the lack of available foraging may lead scout bees from other colonies to smell your entrance feeder and start robbing your hives. Only feed from the top. We use top feeders that hold two jars of sugar water. We add protein powder and other additives to the 1:1 sugar water. The lack of protein in the fall can cause the bees not to prepare their winter brood, which are bees of winter physiology.

Inspecting hives in the fall must be accomplished faster because bees are larger and less patient since foraging is over and there is more honey to protect. Use extra smoke and suit up well. The hive is a full-size, fully packed hive now.

WINTERING THE HIVE

Prepare your hive for winter by providing a windbreak, such as a fence or wall. Many beekeepers prefer to wrap their hives with insulation or roofing paper. Top liquid feeders will not suffice in the winter as bees will not consume cold sugar water. Instead, a candy board is the preferred method of winter feeding. A candy board can be mixed and made or purchased and placed on the hive. A good candy board should also contain protein, vitamins, and minerals. Feeding bees in the winter is the key to winter survival as it allows bees to continue to produce heat based on moving their wing muscles. This movement and activity requires the consumption of much food. Feeding candy boards from the top will allow the queen to continue laying enough brood to keep the hive thriving. The colony needs around 40,000 bees to produce the heat to combat brutally cold weather. Bees do not hibernate; they cluster together around the queen and around developing brood to stay warm.

MULTITASKING IN THE YARD

During the spring, you may consider planting your gardens and flowers near your beehive to maximize pollination. You can also consider moving your hives to your fruit and vegetable gardens. Research and gather information about your growing season, such as when certain plants and flowers start blooming. During your first year as a beekeeper, you will likely not receive your bees until after fruit trees have bloomed. For example, apple trees bloom in our area around the end of March or first of April. Packages of bees are not available until the middle or end of April. However, every year after that, your bees will be able to forage long before any flowers or fruit trees start blooming. If you have the space, consider identifying and planting flowers, trees, and crops that bloom at various times throughout the year. This will provide your bees with the resources they need throughout the seasons. You can plant fruit trees that bloom in the spring, white Dutch clover that flowers throughout the summer, sunflowers that bloom in late summer, and asters such as goldenrod that bloom in the fall. These are just a few examples of options available as you plan your seasonal gardening activities to help support your beekeeping hobby.

WINTER

In northern climates, bees cluster and remain calm but active inside the hive during winter, staying warm by producing their own heat. Winter colonies need 60 to 80 pounds of stored nectar to survive the winter. Excessive moisture in a hive during the winter can be a problem as it freezes at the top of the hive, then melts and drips down onto the winter cluster. Bees can survive cold winters, but not if they are wet. Here are a few tips to keep your hives dry in the winter. Use a screen bottom board and do not close it completely. Provide a slight upper vent on the top of the hive. This will give the bees a way to get humid, stale air out of the hive in the winter. An upper candy board, such as the Winter-Bee-Kind, can absorb excessive moisture into the candy. Do not stack hay bales against the hive as a wind block, because they will hold moisture. Place them several feet away from the hive. Do not feed your bees liquid sugar in the winter, only hard sugar, to help control moisture in the hive.

KEEPING HEALTHY BEES AND HAPPY NEIGHBORS

 hile most people will be thrilled with the idea of a backyard beehive existing in their community, there are things you need to do to keep your bees healthy and, in doing so, keep your neighbors happy with your new hobby.

KEEPING HEALTHY BEES

When you bring in honeybees to your backyard, you are now responsible for keeping the bees healthy. Just as with any livestock or family pet, homeowners must provide a healthy living environment, plenty of water and food (floral sources), and treat any diseases and pests that may be encountered. This is true wherever you live, but it is even more important when you are near homes, businesses, and schools.

Learning how to identify, prevent, and treat these common diseases will help you keep healthy bees. You can find workshops in most states providing training in pest and disease control. Many states have apiary inspectors who can assist you in identifying diseases. Your state may also have college extension agents who specialize in bees and can help you get the information you need.

Varroa mites (*Varroa jacobsoni*). The varroa mite has now been identified as one of the leading causes behind Colony Collapse Disorder and winter die-offs. A small, reddish-brown parasite, it feeds on the blood of bees and the fat bodies of larvae and pupae while vectoring viruses to the honeybee. Methods of prevention include using a green drone comb, breaking the brood cycle, doing powdered sugar dustings, or taking a more aggressive approach, such as using formic or oxalic acid treatments if mite numbers are over acceptable thresholds.

Tracheal mites (*Acarapis woodi*). Tiny, dust speck–sized parasites that live within the breathing tubes or in the air sacs of the bee, these mites can only be found through dissection and microscopic inspection. Symptoms may include bees crawling and unable to fly, disjointed wings (called K wing), and trembling. Treatments include grease patties and menthol.

Nosema (*Nosema apis*). This unicellular parasite is responsible for a disease that produces symptoms of dysentery (diarrhea) in the bee, which may defecate on or inside the beehive rather than outside as healthy bees do. Making sure you have strong hives and replacing aging queens is the best prevention, but hard cases may require the application of the antibiotic fumagillin.

Small hive beetle (*Aethina tumida*). While this is the newest pest, it is now one of the most widespread infestations in beehives. A small black beetle (about ⅓ the size of a bee's body) covered in fine hair, it lays eggs in the honeycomb, where the larvae consume the pollen and comb. Prevent it aggressively with beetle oil traps.

Wax moths (*Galleria mellonella*). This is an adult moth that lays eggs on the comb and in the cracks of hive boxes of weak and dead colonies or in stored boxes. Larvae destroy the comb by chewing through it to find pollen and cast-off cocoons, leaving the hive a mass of webbing and debris. Keeping hives strong and in large numbers is the greatest deterrent to this pest.

Tropilaelaps. This mite is larger than the *Varroa destructor* mite. It is currently not in the United States. Treatment for the varroa mite also treats tropilaelaps simultaneously.

American foulbrood. Caused by the bacteria *Bacillus larvae*, it infects and kills the honeybee larvae. Larvae become dark brown and sunken and have a sour odor. Because spores are highly resistant to treatments and can remain viable for decades within the honey and combs, the only real way to kill it is the burning of infected hives and equipment. Laws vary from state to state on the use of chemical treatments for AFB.

European foulbrood. The *Melissococcus pluton* bacterium causes this disease where symptoms appear as a spotty brood pattern and irregular-colored brood with the larvae twisted in the cell. Aggressive treatment includes **requeening** and using the antibiotic oxytetracycline (Terramycin).

PROPERLY DISPOSING OF WASTE

One of the biggest benefits that a hive provides is abundant honey and wax. Anything inside a hive can be used and reused by humans, greatly reducing waste. **Uncapping** the honey produces large amounts of wax that can be cleaned and melted down to sell or personally used for candles, balms, lotions, or beeswax wraps. Propolis collected from scraping down frames and handles or from traps can be made into tinctures and used for medicinal purposes. Excess pollen can also be collected and used for nutritional purposes. There is little to waste in a hive and in a neighborhood there is little room to waste it, so if you can't use it, wrap it and tightly secure it in plastic, then place it inside a closed can with heavy lid for refuse collection.

DIVIDING YOUR COLONY

Going into your second year of your beekeeper adventure (and in the years following), you will, hopefully and joyfully, find that your colonies have expanded enough to be split into a second beehive. With queens laying upward of 1,500 eggs per day, your hive will grow from its original 3 pounds (10,000 bees) to several more pounds than that by the end of the first summer. You may quickly find yourself with a too-large colony that swarms on a frequent basis and requires much more maintenance. At this point, you will need to make the choice of whether to split your hive.

Some questions you may need to consider about dividing a hive are if you have the room for another hive, if the added work is practical for you, if you are able to use the additional resources of the hive (honey, wax, propolis, pollen), and whether you have the extra time and money needed to maintain a second setup.

If a hive needs dividing, some beekeepers will split the hive, giving away or selling those frames to someone wanting to start beekeeping. If you decide to keep the split, picking up one deep box and walking it away to another location in your yard (called a walk-away split) should be all that you need to do to quickly split the hive. You'll need an additional bottom board, inner cover, and top cover for this new hive, and then additional deep supers as the summer progresses. Depending on where the queen is located now, the other hive should begin the task of making a new one.

TIPS FOR SCALING

» Hives should be at least 70 to 80 pounds in weight, especially going into winter.

» You can manually lift a hive at the back to estimate the weight.

» When using a non-digital bathroom scale, lift one side of your hive to weigh it, multiplying by two to guestimate its current weight.

» Alternatively, you can weigh each box individually, adding the weights together.

» Try using a luggage scale, making sure all boxes are strapped together.

» Measure hives at the same time of day each time.

» Weigh hives every two to three days.

» Weighing tools can be purchased commercially at beehacker.com, beeweigh2.wordpress.com, and broodminder.com.

KEEPING YOUR BEES TO YOURSELF

Coming home from work one day, we saw that a swarm of bees (also known as a bivouac) was dangling precariously from the bar at the top of our son's swing set. A swarm may be only a minor annoyance to us, but it's understandable that a neighbor could become frightened when seeing a ball of stinging insects around their children or pets. Watching swarms land in their trees or on their cars can be alarming, so knowing how to keep your bees from swarming is important in densely populated communities. Spending time to empathize, understand, and rectify any issues will ease the transition to living happily with honeybees when you're close with your neighbors.

SWARM PREVENTION

Swarming is a natural instinct of the bees to split and divide in an effort to increase their numbers. A new queen is created for the hive and, just prior to her emergence, the old queen knows it's time to leave the hive with half of the population. This split emerges in a large cluster, typically first going to a landing site while they await the reports on a new home from the scout bees. Bees swarming off into woods or the forest will increase the feral populations of hives, but in a city, swarms often set up their new hives in homes, churches, or businesses.

To prevent this, a backyard beekeeper will try to keep swarming tendencies to a minimum. Always make sure that hives have additional honey supers ready to go for bees to expand into. Splitting, or dividing a hive either through giving away several frames of bees and brood or making a new hive, decreases the number of bees in a hive, which increases the workload for the bees that are left and eliminates the need to swarm. Requeening a hive that has the propensity to frequently swarm may decrease that tendency.

TROUBLESHOOTING

Understanding any potential problems that can arise will increase your enjoyment of beekeeping and guarantee healthy bees, all while making you a better neighbor. Learning about hive maintenance, inspection techniques, and seasonal management will increase your knowledge and prevent issues. A beekeeper can't avoid all mistakes, but understanding the more common problems will help you be prepared for them.

OVERCROWDING

In the summer, our family hangs out on the deck, lemonade glasses in hand, watching the fireflies as the sun sets. It is common to also see our bees on the outside of a hive in the evenings in the summer, just like us. Inside the hive, the bees are trying to maintain the temperature at 92.5 degrees Fahrenheit, necessary for brood incubation. On a very warm day, bees will congregate, or beard, on the front of the hive to help keep the inside temperature at the right level for larval development.

This is very common, but if you see this behavior on much cooler days, the bees may be indicating that there is a bigger problem with overcrowding inside the hive. Upon inspecting this hive, you may notice that all frames in the supers are full, honey is capped, and that there may not even be any space left for the queen to lay eggs. It's time either to add more boxes, divide your hive, or harvest your honey. Catching this before a swarm starts will keep your population as well as your honey harvest strong.

HIVE SIZE PROBLEMS

Are you using the proper size frames in the right boxes? Using frames that are too shallow for a hive will cause the bees to continue to pull out the wax below the frame. Buy frames from reputable US beekeeping supply companies so that you have standardized frame sizes of greater quality. Buying substandard imported frames can cause size issues as well as issues with broken frames that are not well glued and nailed.

We've mentioned before that overcrowding in a hive can lead to swarming, but the hive can also have the opposite problem of too much space. Add additional supers to a hive only after the bees have filled 6 or 7 frames completely with honey, pollen, and brood. Having too much empty space in a hive can overwhelm the bees, who may wander aimlessly from frame to frame.

Check your queen excluders periodically to make sure they are not bent or pushed too far off the box, allowing the queen to go up into the honey supers and lay eggs. Don't put on your queen excluders until after the bees have begun to start work in the honey supers or the excluders become what the old-timers call "honey excluders." After your bees have begun to work in a honey super, go back in and slip your queen excluder between your brood box and the honey supers, and the bees will continue to walk through it to work the supers.

MISTAKES TO AVOID

Always make sure you have the equipment you need before you need it, including additional supers and the correct size frames. Have on hand an emergency swarm kit that you can use if your hive swarms or is over-crowded. Keep your hive queen-right (meaning you have a healthy laying queen in your hive) so you aren't scrambling at the last second trying to locate a new queen, especially in late fall.

Harvesting too much honey can be bad for a hive that doesn't have sufficient food for winter survival. Heavily feed bees 1:1 sugar water in the fall to help them replenish and build up stores after floral sources are gone, and provide candy boards in the winter. Many beekeepers wait and harvest honey from the hive only in the spring, after they've determined the bees did not need the surplus during the winter.

PART 3

Enjoying the Harvest

One of the most rewarding aspects of backyard beekeeping is the joy of harvesting honey and making use of leftover beeswax from the hive. In this chapter, we will take a look at how to capture honey from specific crops. Honey can have a distinctive flavor based on its floral sources. We will explain how to single out specific flavors and how to harvest, bottle, and market your honey. We'll also share best uses for your leftover beeswax.

HOW TO HARVEST HONEY AND BEESWAX

While you might enjoy the actual task of beekeeping and helping others learn the art, the harvest can be just as exciting. You can cook and bake extensively with honey, which I find gives a unique taste to baked items, keeping them moist longer. You can also use honey for everything from fruit salad to dressing, and use beeswax to make candles, balms, lotions, and food wraps. All of these things may also afford you a small side business where you can meet community members and make new friends, repaying you many times for the original investment in equipment and bees.

WHAT IS HONEY?

How nectar from a flower becomes honey is a fascinating process. Nectar in flowers is designed in every way to entice pollinators with its fragrance and sweetness, tricking the pollinators into picking up pollen and delivering it to nearby flowers. Nectar is comprised mainly of water along with sucrose, glucose, and fructose. After the honeybee visits a flower, she will collect up to 90 percent of her own body weight of the sweet liquid inside her honey stomach, along with pollen. Over the course of her lifetime (about 40 days), a bee will gather about ½ teaspoon of honey. Honeybees fly 55,000 miles and visit two million flowers to bring us one pound of honey. When you consider that a hive can produce in excess of 100 pounds of honey per year, we can only be awed by the sheer number of bees and the vast number of flowers that it took to produce even a single jar of honey.

A bee's honey stomach is a sterile storage tank that also secretes beneficial enzymes (converted into gluconic acid and hydrogen peroxide) into the nectar, which is then forced out of the stomach by muscles and passed off to another bee. This bee will take droplets into her mandibles and will maneuver them back and forth before putting them into prepared honeycomb. Bees will fan the nectar with their wings, where the water content will drop to around 18 percent, at which time the bees will cap it over with beeswax, all sealed up and ready to be eaten, either by them or us.

GETTING THE HONEY YOU WANT

Not all areas of the country yield enough floral sources for honeybees. When bees are not able to bring in enough nectar for their own use, or for the beekeeper's use, steps have to be taken to make sure the bees can access a wide variety of resources. Bees left in monocrops or in monofloral sources, no matter how wonderful, will suffer from poor nutrition and stress, making them more vulnerable to pests and diseases, which greatly affects the rest of the hive as well as the quantity and quality of the honey. Make sure that bees have a wide variety of floral sources as well as fresh water. If you find you have to mainly feed your bees sugar water due to a lack of floral sources, consider moving your hives to a more favorable area. Do not eat this sugar-derived "honey."

Harvesting after a particular flower's *honey flow* (characterized by thousands of bees rapidly going in and out of the hive headed in the same direction) will keep your honey to somewhat singular strains, which may cause the honey to have a specific flavor and coloring. For consumers who prefer a lighter, fruitier honey, harvest in early summer. If a customer prefers a dark, molasses-type honey, extract honey at the end of the fall, when autumn floral sources produce rich, mellow tones. Dark honey is also valuable for sore throats and coughs because it contains a higher amount of antioxidants than does the lighter spring honey. To be labeled a specific kind of honey, samples must be sent for pollen sourcing and contain not less than 51 percent of the pollen from the specific flower.

HARVESTING TECHNIQUES

Backyard beekeepers use the term *harvest* to describe the process of extracting honey, producing raw honey that has not been heated and retains much of the pollen. Commercial companies, on the other hand, often heat their honey to flow more easily through processing machines while heavily filtering out pollen, before it shows up on the store shelves.

Beekeepers know that honey is ready to be harvested when it is capped over by the bees at the right moisture content. Honey that has a too-high moisture content has a very vinegary taste. (Note: *Vinegar honey,* a popular commodity, is produced through a very different process than taking unripe honey.) Harvest frames that are at least 80 to 90 percent capped over, leaving the rest for the bees to finish. If there are other frames that are not 80 to 90 percent capped over, make a note to come back in a few weeks to see if they are ready for harvest. A **refractometer**, an instrument that measures moisture content in foods like honey and jams, and which can be found inexpensively online, can aid beekeepers who need extra reassurance that their honey is ready.

To capture different colors and flavors, harvest capped frames at the end of each season from spring to fall. Beekeepers in cities may need to harvest as often as possible if the heavy weight on roofs, balconies, or other structures is an issue. Some beekeepers choose to wait and harvest only one time per year, and so will harvest in late fall or the following spring. Because honey does not spoil, beekeepers may choose to leave honey supers on the hive for winter food for the bees, harvesting only in the spring when it's determined that the bees did not need it.

Extracting honey varies according to which type of hive you use. With top bar type hives, honey may be pulled out and crushed by hand, necessitating only the need for a bucket and cheesecloth. Hives with honey supers that use turnkey-type technology may also only require a bucket and strainer. Backyard beekeepers typically need only a 2- to 4-frame extractor and should look for well-made, heavy-duty metal canisters with legs, as well as food-grade buckets.

Extracting equipment can be purchased at beekeeping supply companies, farm stores, and online. Many beekeeping clubs, companies, or nature centers will loan out or charge nominal amounts for extracting equipment. Repurpose household items you already own for extracting duty, such as pulp and jam strainers, food buckets, or serrated kitchen knives. Items you may wish to purchase, borrow, or make include:

» Extractor

» Strainer or cheesecloth

» Uncapping knife—either *hot*, meaning you can plug it in and set a thermostat, or *cold*, meaning not electric

» Uncapping fork, or uncapping scratcher

» Uncapping tank or plastic tote with lid for wax cappings

» Food-grade buckets with honey gate

» Food-grade buckets with lids for storage of honey

» Clean, sterile jars with lids for bottling

» A glass of wine, the soundtrack from *Breakfast at Tiffany's*, and a few good friends

LANGSTROTH HIVE

We found our three-year-old grandson dripping in honey one day. He had simply plunged his hand into a frame of honey, and pulling up a chunk of it, was eating it out of his hands and slurping up drips of the goo that were running down his arm. It is possible to get honey out of a Langstroth hive through the *destruct method* (pulling out the comb and crushing it through cheesecloth) like our grandson did, but it is easier, faster, and far less messy to use an extractor. By using an extractor, we are able to retain most of the comb on the frame and then return it to the hive for continued use. And because Langstroth hives are capable of producing anywhere from 50 to 100 pounds or more of honey each year, an extractor is a great time-saving device.

Let's consider first the area that you will use to extract your honey. Most hobbyist beekeepers use their kitchens, but many also set up *honey houses*. The area you choose to extract in should be clean and free of litter, dust, and animal dander. You should refrain from smoking or eating while you extract your honey. Make sure all doors and windows are shut, or you may have honeybees wanting to reclaim their honey! Have access to hot water for cleaning up.

Taking one frame of 80 to 90 percent capped honey, lay a corner of the frame on the lip of a five-gallon bucket. Lay your uncapping knife along the frame edge and slice through the wax *cappings*. Your knife should remain at the level of the frame without digging any deeper into the honeycomb. The cappings will fall into the bucket to be washed and melted later on for other projects like candles and lip balms. With the uncapping scratcher (or kitchen fork), scrape open any unopened cells that you did not get with the knife.

Your frames are ready for the extractor. Smaller extractors are typically for 2 to 4 frames and use a hand crank, but there are larger mechanized extractors that take multiple frames and go to work with the touch of a button. Inside the extractor is a frame basket that is radial or tangential. In radial baskets, the frames sit with the top bar of the frame facing outward. In tangential baskets, one side of the comb faces outward, and after spinning, the frames must be reoriented (or turned) to the other side to be spun again.

When the extractor spins, honey is forced out of the wax by centrifugal force, where the honey hits the side of the basket and drips down to the bottom of the drum. At the bottom of the drum, there is a *honey gate*, which is a valve that can be opened. Under the valve, a bucket is placed topped with a strainer or cheesecloth, and upon opening the honey gate, the honey will flow out into the strainer and into the bucket. Allow the honey to sit for up to several days to allow any bubbles or wax to rise to the top.

TOP BAR HIVE

It is encouraged to not harvest honey from a top bar hive until the spring, allowing the bees to use the comb as thermal protection in winter or for winter food needs. Because the comb in top bar hives is dependent only on the single (generally) bar at the top, these kinds of frames are not suitable for uncapping with knives or for use in extractors. In order to extract honey from this kind of hive, you will have to select appropriate frames, starting from the far end of the hive, that are capped over at least 80 to 90 percent.

Pull the comb off the bar by hand or by slicing with a knife. With a strainer in the top of a food-grade bucket, place the comb into the strainer and crush with your hands, squeezing it gently to pop open the cells to release the honey. Alternatively, wrap the comb in the cheesecloth, twisting shut at the top and squeezing the bag. Allow the honey to drip through the strainer and into the bucket, returning to continue squeezing until all the honey is out of the comb. Keep this comb in a separate bucket or container for further use. If desired, you can also cut the comb into squares, without crushing and straining, and place into sealed containers. Each comb frame will produce up to six pounds of liquid honey.

WARRÉ HIVE

The same *destruct*, or crush-and-strain method, that is used for a top bar hive can be used for a Warré hive that has similar type frames. For Warré hives that use frames with a foundation similar to Langstroth hives, an extractor can be used. Conversely, this comb can also be cut into squares and placed into containers.

Take only capped frames from boxes that no longer contain brood. Warré hive frames contain about eight pounds of honey per frame.

CANNING/JARRING TIPS

Honey is antimicrobial and antibacterial, and as a result, never spoils. All raw honey eventually crystallizes as well, which is what takes place when molecules in the honey form crystals. Various things cause this to happen more quickly, such as the particular kind of nectar the honey is derived from and the temperature. Commercial honey companies heat their honey to kill these crystals, appealing to customers who may think crystallization is spoilage, when in fact it shows the purity of the raw product. To bring crystallized honey back to a liquid state, simmer a pan of water on the stove, placing the jar of honey into the water after it's been removed from the stove.

» Glass jars are preferred by most beekeepers to plastic. Glass is sustainable, recyclable, and easy to clean. Glass can be warmed if honey crystallizes, unlike plastic.

» Consider plastic if you intend to transport or ship large quantities of honey.

» Buy 1-, 2-, and 3-pound glass jars for most customers. Most of your friends and family will also prefer these sizes as well.

» Have available 5- and 10-pound plastic containers for customers who buy in bulk.

» Store honey in a cool, dry, dark cupboard.

SELLING AND MARKETING YOUR HONEY

One of the most exciting and rewarding parts of extracting and bottling honey is having the opportunity to sell it. Most hobbyists find they recoup their investment quickly and easily by selling some of the honey and wax from their hives. Sheri's father, who started beekeeping in his latter retirement years, loved to put his "honey for sale" sign up at the end of his lane and wait for his customers to roll in. The extra money was nice, but he was far more richly rewarded with the companionship and social interaction with his customers.

Where can you sell your honey? Just about anywhere, but always check with your state for their cottage food laws and with your local government on any zoning regulations for small businesses. Besides putting up a sign and selling it from your home, beekeepers sell their honey through

farmers' markets, boutiques, health food stores, town festivals, craft shows, and local grocers who love to sell local food from area farmers. In our modern world of online shopping, creating a website or selling through a site like Amazon or Etsy gives you a new dynamic set of customers who are looking for organic, artisanal, and creatively bottled and boxed jars for personal use and gift giving.

Honey is priced differently throughout the United States, with prices ranging from around $4 in the northern Midwest to $6 per pound in the South to as much as $10 per pound (and higher!) in cities. Lighter honey typically sells for a higher premium than darker honey, and infused or creamed honey will be a dollar or two higher per jar.

Creating and designing a label can be a lot of fun, but there are regulations that require certain information be included. Federal law requires labels to show the common name "honey" on the label along with your contact information and the weight of the product, shown in pounds, ounces, and grams. No ingredient statement is required as honey contains only one ingredient. Other information can also be included on the label, as long as it is a true statement.

BEESWAX

Most of the wax will stay with the hive, but the wax that has been removed by the uncapping process or during the crush-and-strain method of extraction is now a valuable by-product for our use. Beeswax is made by young bees who are around 2 to 3 weeks old. These workers have wax glands on their undersides that secrete wax scales, which are then taken by other bees who render it pliable enough to form into honeycomb, which holds the brood or honey. These wax scales also become the cappings over finished honey. For bees to produce one pound of wax, they must consume 11 pounds of nectar or honey.

HOW IS IT HARVESTED?

One to several pounds of wax can be harvested from each super. As you are extracting the honey from the wax, collect the wax cappings or the crushed wax in a separate container or bucket. Allow any residual honey to drain out of the wax by holding it in a strainer over a bucket for several days. Run water over the wax to clean it and allow it to dry. Using a dedicated pan or slow cooker, melt down the wax, then run it through another strainer or old T-shirt to remove any honey or debris. While it is still liquid, pour the wax into molds, muffin tins, or metal food cans and allow to harden, storing in plastic bags or containers away from heat.

CHAPTER 10

RECIPES FOR YOUR HONEY

bout 10 years ago, I taught a "cooking with honey" class for a large group of mostly bee-keepers at our training center in Illinois. I asked the group how many cooked or baked with honey, and I was astonished to find out that nearly none of them did. When I asked what they did with their honey, they mentioned spreading it over toast, spooning it into their coffee, and drizzling it over ice cream. An hour later, after demonstrating how to make smoothies, shrimp cooked in a spicy honey sauce, and cherry cheesecake—all with honey—I had made new converts.

I started cooking and baking with honey nearly 25 years ago when David and I had such a surplus that I needed to do something with all of it. We weren't into the selling avenue of our business yet, and I had already given away all the honey that I could. I knew honey was better for us than sugar, but I was far more interested in learning to use it in my own kitchen and in turn save a few dollars on the weekly grocery shopping. And the great thing was that I learned I could use honey to replace sugar in nearly everything I made, from iced tea to marinades to basting sauces, as well as in my lusted-after homemade breads and cookies. Not only did the honey give a uniquely mellow, sweet taste to my recipes, I found I could use less sweetener (honey is sweeter than sugar) and the items stayed moist longer.

Just a few tips will help you convert any recipe into a honey favorite, but I rely mostly on recipes online and in cookbooks that have already made the conversions for me. When cooking or in marinades, dressings, and sauces, honey can be equally substituted for sugar. In baking, reduce the recipe for every cup of sugar to only ⅔ cup of honey. For every 1 cup of honey, subtract ¼ cup of another liquid in your recipe, and add ¼ teaspoon baking soda for every 1 cup if your recipe doesn't include it (because honey is slightly acidic). Reduce baking temperatures by 25 degrees as honey makes the baked goods brown more quickly.

CHILE-INFUSED HONEY

Makes 1 cup

My chile-infused honey is a big hit with everyone. I can make it as sweet-mild or spicy-hot as we want. Then I put it in plastic ketchup-type bottles and mark it "mild," "medium," or "hot," and use it over shredded meats, ribs, chicken, and even pizza!

4 dried chiles or chipotle peppers

1 cup honey

1. In a blender or food processor, pulse about 4 dried chile peppers. (You can use more or less to personal taste, and you can substitute other types of peppers.)

2. Heat 1 cup of honey to around 100°F. Do not go any higher than this or you may kill some of the beneficial enzymes in the honey. Add the chile peppers, continuing to stir for about 15 minutes. Let cool for 5 minutes.

3. Strain out the chile peppers. Reserve 1/2 cup of the mixture for garnishing.

HONEY BARBECUE SAUCE

Makes about 24 ounces

In my barbeque sauces, I use chile-infused honey in place of brown sugar or molasses and spices. Great not only for shredded pork, beef, or chicken, it's my go-to sauce to use with navy beans to whip up a batch of Boston baked beans.

½ cup Chile-Infused Honey (page 121)

1½ cups ketchup

½ cup red wine vinegar

½ cup water

1 tablespoon Worcestershire sauce

2½ tablespoons dry mustard

2 teaspoons paprika

2 teaspoons salt

½ teaspoon pepper

1. In a blender, combine chile-infused honey, ketchup, vinegar, water, and Worcestershire sauce.

2. Season with mustard, paprika, salt, and pepper, blending until smooth.

3. In saucepan, heat on low, stirring occasionally, allowing flavors to meld, about 10 minutes.

4. Spread on ribs on outdoor cookers, stir into pulled pork or shredded beef and chicken, or put into ketchup-type squeeze bottle, as a condiment.

HONEY COOKIES

Makes 2 to 3 dozen large cookies, or 4 dozen small cookies

Our daughter held a honeybee-themed birthday party for our granddaughter and made these delicious honey cookies.

½ cup shortening

1 cup creamy peanut butter

1 cup honey

2 eggs, slightly beaten

3 cups flour

1 cup sugar

1½ teaspoons baking soda

½ teaspoon salt

Optional topping: 1 cup chocolate chips

1. Preheat the oven to 350°F.

2. In a mixing bowl, cream together the shortening, peanut butter, and honey.

3. Add eggs one at a time, mixing well.

4. Combine the flour, sugar, baking soda, and salt in a separate bowl.

5. Add the dry ingredients to the peanut butter and honey mixture and incorporate well.

6. Roll into 1- to 1½-inch balls and place on ungreased baking sheets. Flatten with a fork dipped in sugar. Bake for 8 to 10 minutes and let cool completely.

7. Optional topping: Microwave chips on medium power for 1 minute, then stir. Continue heating the chocolate at 10- to 15-second intervals, stirring between each one, until it is almost melted.

8. Squeeze chocolate into forked lines to represent bees.

SHERI'S HONEY-BRAIDED HOLIDAY BREAD

Makes 2 small loaves, or 1 large braided loaf

I make these braided loaves in batches of 10 or more for holiday gifts (yes, they are that wonderful!) and found the workload is lightened by using a stand mixer or bread machine for kneading.

1 (¼-ounce) package active dry yeast

1 cup plus 2 tablespoons warm milk (110° to 115°F)

¼ cup butter, melted

¼ cup honey

1 egg

1 teaspoon salt

4 cups all-purpose flour, divided (plus ¼ cup extra on hand in case the bread is too sticky)

Egg wash:

1 egg, beaten

1. Preheat the oven to 350°F.

2. In a large mixing bowl, dissolve the yeast in the warm milk.

3. Add the butter, honey, egg, salt, and 2 cups flour; beat until smooth.

4. Stir in the remaining flour to form a soft dough.

5. Turn onto a floured surface and knead until smooth, about 6 to 8 minutes.

6. Place in a greased bowl and turn once. Cover and let rise until doubled, about 1 hour.

7. Punch the dough down and then turn out on a floured surface. Divide the dough into thirds, shaping it into a rope or wreath by braiding.

8. Cover and let rise until doubled, about 45 minutes. Brush with egg wash. Bake on a greased cookie sheet for 20 to 25 minutes until golden brown.

CHERRY ALMOND HONEY GRANOLA

Granola is full of protein, fiber, and nutrients. Make it even healthier by substituting out the sugar for honey.

3 cups rolled oats

1 cup sliced almonds, roughly chopped

½ cup walnuts

¼ cup flaxseed

¼ cup sunflower seeds

¼ cup brown sugar

¼ cup plus 2 tablespoons honey, divided

¼ cup coconut oil

¾ teaspoon salt

¼ teaspoon ground cinnamon

¼ teaspoon nutmeg

1 teaspoon vanilla extract

Dried cherries, amount to your liking

1. Preheat the oven to 250°F.

2. Line a large rimmed sheet pan with parchment or wax paper. Set aside.

3. In a large bowl combine the oats, almonds, walnuts, flaxseed, sunflower seeds, and brown sugar. Mix until well combined.

4. In a medium bowl, add ¼ cup of the divided honey, the oil, salt, cinnamon, nutmeg, and vanilla. Hand whisk or use a mixer to make into a syrup. Pour the honey syrup over the dry oat mixture and stir for 3 to 5 minutes or until all dry ingredients are coated and moist.

5. Spread the mixture out evenly in the prepared sheet pan. Drizzle the remaining 2 tablespoons of honey over the top of the granola.

6. Place the sheet pan in the oven and bake for 60 to 75 minutes, stirring every 15 minutes until granola reaches a uniform toasted color.

7. Cool completely, then add in the dried cherries or other dried fruits to your liking.

TIP: Store prepared granola in an airtight container at room temperature for a few weeks or freeze in a zip-top bag for several months.

HONEY YOGURT FRUIT SALAD

Makes about 8 servings (double or triple as needed)

We have a very large family, ranging in age from our 2-year-old grandson to my mother who is 82, so I needed a crowd-pleasing dish that was easy to prepare. My Honey Yogurt Fruit Salad fills the bill.

2 each of any or all: bananas, strawberries, blueberries, mangos, mandarin oranges, apples, grapes

1 orange, juiced and zested

2 cups yogurt (unsweetened)

¼ cup honey

½ teaspoon vanilla extract

1. Slice the fruit and toss with the juice of one orange (to keep the fruit from browning).

2. Stir together the yogurt, honey, and vanilla.

3. Add the fruit and stir. Garnish with orange zest.

FUN TIP: If you add crushed ice, you can blend up this recipe for a really delicious smoothie for any time of day.

HONEY MUSTARD DRESSING

A women's magazine came out to do a feature on our farm, and, during the photography session, I fed everyone green summer salads with these amazing dressings to choose from. Even though we were busy that day from sunup to sundown, it only took a few minutes to put together, just perfect for those really hectic days. You only need four items for salad dressing—oil, acid, spices, and a sweetener—so feel free to change it up.

¼ cup mayonnaise

1 tablespoon store-bought brown mustard

2 tablespoons honey

½ teaspoon lemon juice

Zest of 1 lemon

1. Place all the ingredients in a pint jar with a lid.

2. Shake and use.

3. Refrigerate after use. Allow to come to room temperature and shake well before use.

HONEY POPPY SEED DRESSING

Makes 2 to 4 servings

One of David's favorite salad dressings is one I enjoy making for him. The combination of olive oil, honey, vinegar, poppy seeds, paprika, and Worcestershire sauce in this dressing gives a balance of sweet, sour, and healthy flavor to any salad. Try different types of vinegar to make this recipe unique each time you serve it.

½ cup olive oil

½ cup honey

¼ cup white wine vinegar

1 tablespoon poppy seeds

¼ teaspoon paprika

¼ teaspoon Worcestershire sauce

1. Place all the ingredients in a pint jar with a lid.

2. Shake and use.

3. Refrigerate after use. Allow to come to room temperature and shake well before use.

HONEY ORANGE DRESSING

Makes 2 to 4 servings

My favorite container in which to make salad dressings is just a simple canning jar with a lid. Add all the ingredients and shake away. Store in the refrigerator, but allow it to come to room temperature and shake well again before use.

**2 tablespoons
orange juice**

**2 teaspoons
chopped
green onion**

½ cup honey

½ cup olive oil

Salt

**Freshly ground
black pepper**

1. Place all the ingredients in a pint jar with a lid.

2. Shake and use.

3. Refrigerate after use. Allow to come to room temperature and shake well before use.

HONEY-GLAZED CARROTS

Makes 4 to 6 servings

Carrots provide vitamin A, which is known to improve our vision. Why not make them a bit more tasty using honey, olive oil, mustard, and salt and pepper?

2 bunches of carrots (10 to 13 carrots)

2 tablespoons olive oil

Salt

Freshly ground black pepper

3 tablespoons honey

2 tablespoons stone-ground mustard

1. Preheat the oven to 400°F.
2. Wash and cut off the greens from the carrots. Place the carrots on a baking sheet.
3. Drizzle the olive oil over the carrots, rubbing the oil into the carrots with your hand.
4. Season with salt and pepper.
5. In a small bowl, whisk together the honey and mustard.
6. Drizzle half of the mixture over the carrots.
7. Bake for 15 to 20 minutes, or until tender.
8. Remove from the oven and drizzle with the remaining honey-mustard mixture.

HONEY LEMON GARLIC SALMON

Makes 2 servings

David is something of a salmon connoisseur who loves wild-caught salmon from the Pacific Northwest. This is his favorite recipe.

2 tablespoons olive oil

4 salmon fillets, several ounces each

Salt

Freshly ground black pepper

3 teaspoons garlic, minced

2 tablespoons lemon juice

¼ cup soy sauce

¼ cup honey

Zest of 1 lemon

1½ tablespoons cilantro

1. Heat the olive oil in a large pan over medium heat.

2. Season the salmon with salt and pepper to taste, and place in the pan to cook for 4 to 5 minutes per side or until cooked through.

3. Remove the salmon from the pan, and add the garlic to the pan to sauté for 30 seconds.

4. Add the lemon juice, soy sauce, and honey to the pan; bring to a simmer.

5. Return the salmon to the pan and spoon the sauce over the top until heated through. Serve and garnish with lemon zest and cilantro.

HONEY BERRY WINE SLUSHIES

Serves 2 to 4

Thanksgiving time around here means it's time to bring out the punch bowl so we can make Honey Berry Wine Slushies. Make sure you put an "Adults Only" sign on it!

1 bag frozen berries

1 cup honey

1 bottle of wine, such as rosé, Moscato, Riesling, or chardonnay

Ice

1. Put the berries and honey in a blender and pulse for 30 seconds.

2. Add half the bottle of wine, pulse for 30 seconds, add the remaining wine, and pulse again for 30 seconds.

3. Remove half the mixture, and fill the blender with ice. Blend and pour the wine and ice mixture into a punch bowl.

4. Return the remainder of the wine mixture to the blender and fill again with ice. Blend and add to the punch bowl.

FUN TIP: You can add more wine and less ice and have a very merry holiday.

HONEY SIMPLE SYRUP

Makes 1 cup

Make a simple syrup with honey to add to your favorite cocktails. This syrup can also be used in lemonades, iced tea, coffee, marinades, and vinaigrettes. Store in your refrigerator for up to 1 month.

½ cup honey

½ cup water

1. Combine the honey and water in a small saucepan over medium heat.

2. Stir until blended.

3. Strain into a jar and seal tightly. Will keep for 1 month in refrigerator.

BLACK BASIL WITH HONEY SIMPLE SYRUP

The bright flavors of basil and lemon and the sweetness of honey make this cocktail perfect for a summer evening on the deck.

Ice cubes

1 part vodka

1 part lemon juice

½ part limoncello

½ part Honey Simple Syrup (page 133)

2 basil leaves

Tonic water

1. Fill a shaker with ice.

2. Add the vodka, lemon juice, limoncello, Honey Simple Syrup, and basil.

3. Shake and strain into a glass.

4. Top with tonic water.

COFFEE TONIC WITH HONEY SIMPLE SYRUP

We always have lots of leftover coffee to make this drink later in the day. You can also use different fruit juice to change up the flavor.

1 ounce cold-brew coffee

½ ounce fresh grapefruit juice

½ ounce Honey Simple Syrup (page 133)

¼ ounce fresh ginger juice

Ice

Splash soda water

1. Combine the cold-brew coffee, grapefruit juice, Honey Simple Syrup, and ginger juice in a shaker.

2. Shake until well combined.

3. Strain into a glass over ice.

4. Top with a splash of soda water.

CREATIVE WAYS TO USE BEESWAX

esides honey, beeswax can also be harvested from the hive. Beekeepers who keep bees in Langstroth hives don't generally take more than just the wax cappings, because it is hard work for the bees to make the wax, so we return most of the frames of wax back to the bees for their own use. In top bar and Warré hives, beekeepers may take the entire frame of wax, but do so with the understanding that they should leave plenty behind for the bees.

The beeswax is an incubator for eggs, a cradle for the larvae, and a storage facility for the bees' food. Along with propolis, beeswax is part of the bee's immune system that helps it stay healthy, as well as providing thermal insulation from the cold. Humans, too, have found ways to benefit from the wax of the beehive, making the process of keeping bees sustainable, renewable, and recyclable. Beeswax is antibacterial, antimicrobial, anti-inflammatory, and moisturizing, making it naturally healthy and beneficial for us.

To render beeswax for use in recipes or projects, start with clean beeswax. If you aren't using your own that you have washed out and drained, purchase only local, pre-filtered organic beeswax in blocks or pellets. Grate or cut beeswax into smaller chunks and, with a dedicated boiler, pot, or electric skillet, melt the beeswax on low heat, straining if necessary through cheesecloth or a discarded (clean) T-shirt. Buy tins and pots for your products, candlewicks, and molds from your hobby craft store or online.

BALMS AND SALVES

Whether it's balms for your lips or salves for your feet, these products are virtually identical, easy to make, and very versatile. One of the best things you can do for your lips in harsh weather or in the sun is to hydrate and protect them with a beeswax balm. Aching, tired, and calloused feet? Beeswax salve can moisturize, soothe, and smooth out the roughness. By adding olive oil to beeswax, along with essential oils, you can create balms and salves for yourself, for holiday gifts, or to sell along with your honey. (Customers who may not particularly enjoy honey may love to buy your balms and salves instead!)

For every tablespoon of melted beeswax, add 1 tablespoon of oil (olive, almond, and coconut are my favorites) along with a few drops of your favorite food-grade essential oils. Try grapefruit or lime essential oils; sweet orange, peppermint, or lavender essential oils are big favorites. Adding a few capsules of vitamin E, shea butter, or cocoa butter will add natural SPF properties. Pour into tins and containers and store in the refrigerator, as there are no preservatives in these natural balms.

HAIR POMADE AND MUSTACHE WAX

All the men in my family have beards. After watching some of those beards become rather unruly, I experimented one weekend and discovered I could make some fantastic sculpting wax. A little of this wax rubbed between your fingers to loosen and warm, and then worked through your hair can be used to sculpt and style some types of hair, beards, and mustaches.

It's easy to make. You'll need beeswax, shea butter, jojoba oil, and essential oils. For every ounce of beeswax, add 1.5 ounces of shea butter, 2 ounces of jojoba oil, and around ¼ teaspoon of essential oils. Lime, basil, and lemongrass essential oils can help with oily hair, and lavender and rosemary are helpful for dry hair. Try tea tree oil for dandruff.

After the wax and butter are melted, stir in the jojoba oil. Let the mixture cool slightly before adding the essential oils. Pour into tins and gift to loved ones.

HAIR CLAY

Beeswax, along with other hair moisturizers and natural plant oils, is the foundation of hair clays, which work like pomades. Hair clay includes actual clays like bentonite or kaolin, which can have nourishing effects on some types of hair, detangling while making it soft and manageable.

Clay is a thick, stiffer, grainier type of pomade (because it contains more beeswax than pomades) that is used to loosen curls or to achieve a sleek, slicked-down hairstyle. Hair clay adds volume because it expands as water is added to it and provides long-term nourishment to your scalp and hair. Only use a dime-size amount and keep a comb handy to rework this versatile, flexible look as often as you want during the day.

Hair clay is a commercial salon product that can be purchased online. Source your hair clay from a reputable, cruelty-free company that uses only natural ingredients and extracts.

HAIR COLOR WAX

Do you like to change up your hair color frequently or just glam it up for a special party but don't want to continually color your hair with harsh ammonia products like salon dyes? If you're the parent of a teenager, like we are, you may also want to help your teenager try out a new trend with a product that is ultrasafe, nontoxic, and short-lived.

Hair color wax is just one of the latest trends that is not only a temporary color that washes out in a few shampoos, but is easy to apply with just your fingers, unlike salon-quality dyes. And you don't have to worry about any harsh ingredients in the product.

Use it on just-washed damp hair, evenly distributing it throughout your hair. If you want more vivid color, wait 20 minutes before applying a second layer to intensify the look. It's so easy and safe to use that you can change your color daily! Making your own hair color can be a little messy, so we recommend buying this online.

BEESWAX CANDLES

Pure beeswax candles are a natural product that burn cleaner, longer, and brighter. Burning beeswax can actually purify the air, making these candles a good choice for people with allergies or sinus issues. Beeswax candles have no added fragrances that could possibly bother family and friends, instead getting their scent from the honey and comb, which may subtly change depending on the flower nectar it was derived from. Historically sought-after by royalty and the church, beeswax has become easily obtainable and usable by everyone in our modern age.

Melt down pure beeswax and directly pour into canning or decorative jars, make into pillars or tapers, or fill fancy purchased candle molds. Choose a 60-ply wick and place the end of the wick down into your jar or mold. Using a skewer, push the wick down into place at the bottom of the jar, adding an inch of wax and holding it in place while the wax hardens. Wrap the wick around the skewer and hang off the side of the jar. Pour the remaining wax into the jar or mold, filling the container up to the top. Balance out the skewer with the wick, centering it over the jar as the wax hardens. Place the candle jar into warm water, which should prevent the wax from cracking (if that sort of thing bothers you!) as it hardens. Burn new candles one hour to begin with, and remelt and reuse any unburned candle wax to make new candles.

OLD-FASHIONED BEESWAX SOAP

Gorgeous soaps can be made from beeswax. Soaps can come in brilliant colors, designed and cut in amazing ways, and wrapped in creative papers and labels. I leave my soap in a loaf shape and have customers tell me how many inches they want cut from it for them. I wrap the cut pieces in wax paper, and my customers really love it.

For many centuries, soap has been made from beeswax and lye, which is a by-product of wood ashes. Recipes can be found online for this kind of soap, but lye is tricky and hazardous to use, causing chemical burns if it comes into contact with skin. Using a saponified soap base (melt and pour) that contains lye is a simpler, less caustic way to make your own soaps. Follow the directions on the package, adding about 1 percent (of the total volume) of beeswax, along with 1 teaspoon of honey per pound of soap. Mixing with a stick blender, add essential oils or colorants and pour into a silicone loaf mold or individual soap molds, following the directions for letting the bar set up properly.

Soaps are great for practical use, gift giving, and as an added-value product to sell in your honey store, at the craft show, or as a boutique item. Teaching a soapmaking class could also be a fun and sociable way to make a few dollars. This is also a fun activity if you need a project to do with your kids, scouts, or a youth group.

BEESWAX FOOD WRAPS

A product that has exciting uses is beeswax food wraps. While not a new concept, there has been renewed interest in these wraps. Beeswax is melted along with jojoba oil and pine resin and applied with a brush to 100 percent cotton fabric, which has been cut into squares of various sizes. After applying the wax and allowing the squares to dry, these wraps are suitable for sealing jars or for wrapping items for refrigeration in lieu of plastic wrap or aluminum foil. This cuts down on garbage waste, saves money, and thus is better for the environment by using a reusable product.

The wax is softened by the warmth of your hands as you mold it to the jar or food item. Wash after each use, thoroughly let dry, and reuse over and over for about one year, depending on the frequency of use. These wraps are not recommended to be used for meats or cheese.

Source pine resin and oil from reputable companies. These items are also good not only for personal use but for gifts or to add alongside your other beeswax items in your own store.

FIRST AID: HEALING HERBAL SALVES

You can easily make a salve to deal with minor first aid problems, rashes, bites, and stings by infusing healing herbs into beeswax and oil. Throw some in the diaper bag for baby bottom rashes, rub into torn cuticles, or smooth over psoriasis and eczema. Cuts, scrapes, and sunburns can all be calmed with a naturally made beeswax herbal salve.

A healing salve starts with infusing the herbs. Fill a jar with herbs and cover with oil, leaving about an inch at the top of the jar. Place a lid on the jar and let it sit for about 4 to 6 weeks, occasionally shaking the jar. At the end of the infusion time, strain out the material from the oil, discarding the herbs. Popular herbs used for healing purposes include comfrey, chickweed, calendula, lavender, meadowsweet, frankincense, and goldenseal. Research the ones that will work the best for you and your family.

Next, heat 1 ounce beeswax with 1 cup coconut or olive oil and 1 tablespoon vitamin E for antioxidant effects and skin benefits. If you desire a creamier salve, add more oil or lanolin. Add the infused herbs. Allow to cure for one week and store in the refrigerator for up to one year. Keep in the cooler when you are out enjoying nature and on vacation.

CRAYONS

For true do-it-yourselfers and crafters, beeswax crayons can be another fun item to make and one that'll make the kids happy as well.

Traditional crayons are made with paraffin, which is a by-product of crude oil and does not decompose, filling landfills with crayon stubs. Other ingredients in crayons may include talc fillers that can be contaminated with toxic fibers. The great news is that crayons made from beeswax are completely biodegradable and better for the environment as well as nontoxic to your little ones.

Beeswax, cocoa butter, and carnauba wax are melted together and mixed with pigments before being poured into crayon molds (although it's even more fun to use molds of all sorts of shapes and sizes!). Because this project calls for hot wax and butter, it is not recommended to do them around small children. These crayons are a great gift idea as well as another idea for the boutique and craft seller.

MODELING CLAY/BEESWAX

Crayons aren't the only thing for the kids you can make. Did you know you can make a very simple and easy modeling clay using beeswax? Start by melting ½ cup beeswax and then adding 1 teaspoon jojoba oil and ½ teaspoon lanolin oil. Next add the colorant and pour the melted wax into paper-lined muffin cups. Continue this with each of the different colors you want. Remove the paper as soon as the modeling beeswax cools down enough to handle, but don't let it completely harden or it may be hard to remove the paper.

Modeling beeswax will be of a harder consistency than regular modeling clay after it cools and sets. To use, break off a small chunk and soften it by holding it in your fist. Alternatively, you can set it on a heating pad to warm it up enough to be pliable. I like to put out wax paper for the kids to use the modeling beeswax on, because it will leave greasy streaks on a table or counter.

The really great thing about modeling beeswax is that it doesn't dry out. The kids can mold artwork into permanent pieces or easily dismantle their wonderful sculptures to reuse the modeling beeswax over and over. Store indefinitely in plastic bags or containers.

FURNITURE POLISH

Modern wood polishes can contain solvents and chemicals, especially those that are derived from the refining and processing of petroleum. A good, natural alternative for furniture polish is beeswax furniture wax.

Beeswax furniture polish is made from melting down ¼ cup beeswax and mixing with ¾ cup olive or coconut oil. You can also add several drops of an antioxidant like vitamin E, which will help prevent the oil from going rancid. Place the polish into a widemouthed jar or tin. After cooling, it's easy to apply by just scooping some out with a rag. Polish your wood piece or furniture in a circular motion, being mindful not to overapply in crevices and moldings. Wipe off any excess oil, then buff it by going over it again with a clean cloth, pulling out any excess wax from cracks.

Use this great-smelling beeswax polish to clean up damaged pieces, to cover and fill scratches, or as a finisher coat on unfinished wood. Beeswax will restore the shine and gloss to your furniture and leave behind a natural, beautiful beeswax-and-honey scent instead of artificial and harmful synthetic fragrances. Soften hardened furniture wax by placing in the microwave for 10-second intervals, stirring and repeating until it begins to soften. Will keep indefinitely stored in a cool, dry area.

WATERPROOFING

For centuries, people have waterproofed clothing, bags, sheaths, boots, and shoes with beeswax, which is still a near-perfect way to keep out the cold and wet. Even with our modern synthetic fibers, not everything is waterproof. To waterproof leather items, paint or brush a layer of melted wax onto the leather, making sure to work it into the leather well. Using a hair dryer, heat up the wax on the boots, where it will seep into the leather and become a barrier against the weather. Other fabrics like canvas or tapestries can be waterproofed as well, but note that it will make the materials darker.

Wood can also be waterproofed. Did you know that although most beekeepers choose to paint their beehives, many often brush melted beeswax onto their hive boxes to protect them? If the wax cools down too fast while you are applying it, you can take a blower or heater to melt down the wax again so that it seeps inside the wood, making the box waterproofed and protecting it against the elements. You can also use beeswax this way on any unfinished outdoor furniture.

Other items that can be waterproofed include motorcycle leathers, saddles, tack, and chaps. Wax can be used on gloves, briefcases, even suede items. Most items will need to be re-waxed at least once a year or more if the items are used heavily in the weather.

LUBRICANTS, SEALANTS, AND RUST PREVENTION

Beeswax is also an excellent lubricating compound for items around the house like window frames, drawer slides, and zippers. Do you have something that is stuck? Just give it a rub of beeswax to make these items move, glide, and shut easily. Rub wax over the threads of screws and nails to help resist corrosion and to make them drive more smoothly. Use wax on the flat side of saw blades, which keeps them sharper longer by reducing wear on the teeth.

Seal exposed beams in your home with brushed-on melted wax, while rubbing wax on copper or concrete both seals and preserves and will possess a high shine when buffed. Blacksmiths use a mixture of beeswax, linseed oil, and mineral spirits to prevent rust on the tools they use. Heated forged items are rubbed over with wax that melts and seeps into the pores of the metal, sealing against corrosion. After buffing, the wax sealant gives the metal a rich, warm glow.

SURFBOARD WAX

I recently traveled to San Francisco to a food show and chocolatier convention where my daughter and I decided to take in the sights. It was late January, wet, foggy, and very chilly, but the die-hard surfers were out under the Golden Gate Bridge, and with Alcatraz in the background, we stood there in the freezing drizzle and watched those surfers, marveling at their passion and zeal.

Surfers need grip control to prevent slipping off their boards. A terrific surfboard wax can be made from beeswax, pine resin, and coconut oil. There are several different wax types for surfboards: cold, cool, warm, and tropical, and it's essential to pick the right one for the area you are surfing in. Starting with a clean board, apply the wax gently, especially in the foot area. The beeswax will form small beads that will provide better traction and grip.

And the next time you land that perfect radical layback snap, you can thank the honeybees.

MISCELLANEOUS USES

Beeswax has been used in creating art from the beginning of time. Batik has been an interest of mine lately, although it has quite a learning curve. By using a tool called a *tjanting*, which is filled with melted beeswax, a design can be drawn, printed, or stamped on fabric, which is then dyed and dried. The wax is then peeled from the fabric, leaving uncolored areas on the material, which is used for highly decorative quilting, curtains, scarves, aprons, and more. Ukrainian *pysanky* eggs are also designed with a hot beeswax dispensing pen, using the wax-resist and dye method to make spectacularly colored and inscribed Easter eggs.

Quilters like beeswax to run their needles through, and archers use the wax for bowstrings. Ammo reloaders use beeswax as a powder or bullet lubricant and for grip control just as rock climbers and other athletes do.

Beekeepers use wax to coat their foundations on beehive frames, by brushing on melted wax. This gives the bees a head start in building up their honeycomb in a new hive. Beekeepers also use swarm lure phero-mone stirred into melted beeswax that is solidified into bars and then rubbed all over the inside of a beehive in hopes of attracting a wild swarm of bees.

These are just a few of the more than dozens of other uses for beeswax, a natural, healthy, versatile by-product of the hive.

GLOSSARY

absconding: When the entire colony abandons the hive; this can happen for a variety of reasons.

apiary: Where honeybees are kept.

attendant bees: Worker bees that feed and groom the queen.

bearding: When a large number of bees crawl outside their hive and gather on the entrance or in front of the hive in order to alleviate some internal heat.

bee bread: A mixture of honey, pollen, nectar, and bee saliva that creates a stored, fermenting food source for the colony.

beeswax: The secretion made by young bees who have wax glands on their undersides; used to form the honeycomb, which holds the brood or honey.

brood: A term that refers collectively to the eggs, larvae, and pupae of the honeybees.

capped cells: Cells filled with brood or honey.

colony: All of the bees that live together, including the queen, drones, workers, eggs, larvae, and pupae.

colony collapse disorder: An unexplained phenomenon when a majority of bees in a hive disappear.

comb: Six-sided cells made of beeswax where bees store honey and pollen and raise new bees.

dearth: A condition in which there are few to no available floral sources for bees to forage.

drawn comb: Cells built out from a foundation in a frame.

drones: Male bees without stingers whose only role is to mate with new, virgin queens from other colonies.

egg: Small, white, and shaped like rice; bee eggs are found in the base of a cell of honeycomb.

entrance reducer: A block of wood that reduces the size of the entrance to the hive.

established colony: When a hive is completely operational and is (typically) operating at maximum population.

extractor: A machine that extracts honey by spinning the frames.

forager: A worker bee that gathers nectar and pollen.

foundations: Wax forms in the frame of a hive on which bees build their comb.

frame: Racks in the hive on which the bees make honeycomb.

hive: A bee-created or human-made structure, used to house bees.

hive body: The main part of the hive where the bees live.

hive tool: A type of pry bar that the beekeeper uses to pull apart the hive components.

honey bound: A condition resulting from an abundance of foraging that has filled most frames with nectar or honey, limiting the area for brood or pollen.

larva/larvae: A legless and featureless white grub.

laying workers: In the absences of the queen and/or brood

pheromones, the workers begin laying unfertilized eggs that mature into male drones.

moisture level: The percentage of moisture in honey. Honey is best harvested when the moisture is around 18 percent.

nectar: A sweet liquid produced by flowering plants and collected by honeybees.

nuc: A small hive that is already established and includes wax, brood, pollen, honey, and bees (i.e., drones, workers, and a queen).

pheromones: Chemicals that bees produce that are used to communicate with one another.

pollen: Powder-like substance produced by the male parts of a flower.

pollination: The term for fertilization of plants; when pollen is transferred from a male plant to a female plant.

propolis: A sticky substance gathered from plants and trees that is used to strengthen the comb, cement the comb in the hive, and control pathogens.

pupa/pupae: The stage of metamorphosis in which the tiny bee begins to develop its adult body parts; a larva enters a cocoon in order to transform into a pupa.

queen: The only egg-laying female in the colony.

queen excluder: A divider made of material with spacing that allows workers to pass but excludes the queen. It is used below a honey super to prevent the queen from laying eggs in the super.

quilt box: A box for the top of the hive that absorbs moisture from condensation during cold months; it keeps the bees dry and warm.

refractometer: An instrument used to measure the moisture content of honey.

requeening: The act of removing the old queen and replacing her with a new one.

royal jelly: A substance secreted from the hypopharyngeal glands of nurse bees and fed to young developing larvae.

smoker: A device that produces smoke and is used to calm the bees.

super: A hive box added to an established stack of hive boxes.

swarm: When half the bee colony leaves with the old queen to form a new nesting site.

uncapping: The act of removing the beeswax cap of sealed comb cells.

waggle dance: When scout bees vibrate their bodies to indicate the distance or energy that bees must put into their flight to find food.

workers: Bees that are responsible for housekeeping, taking care of the queen and brood, and foraging; these bees are always female.

RESOURCES

Books

Caron, Dewey. *Honey Bee Biology and Beekeeping*. Kalamazoo, Mich.: Wicwas Press, 2013.

Horn, Tammy. *Bees in America: How the Honey Bee Shaped a Nation*. Lexington: University Press of Kentucky, 2006.

Sanford, Malcolm T., and Richard E. Bonney. *Storey's Guide to Keeping Honey Bees*. North Adams, Mass.: Storey Publishing, 2018.

Beekeeping Organizations

American Beekeeping Federation: abfnet.org

American Honey Producers Association: AmericanHoneyProducers.org

Eastern Apicultural Society: EasternApiculture.org

Heartland Apicultural Society: HeartlandBees.org

National Honey Board: Honey.com

Western Apicultural Society: WesternApiculturalSociety.org

Magazines

American Bee Journal: AmericanBeeJournal.com

Bee Culture: BeeCulture.com

Blogs

BasicBeekeeping.blogspot.com

BeekeepingLikeAGirl.com

ScientificBeekeeping.com

Online Tools and Apps

Beescape.org

Broodminder.com

Pollinator.org

REFERENCES

Chapter One
"Helping Agriculture's Helpful Honey Bees." U.S. Food and Drug Administration. Accessed March 11, 2020. FDA.gov/animal-veterinary/animal-health-literacy/helping-agricultures-helpful-honey-bees.

Lusby, Patricia E., Alexandra L. Coombes, and Jenny M. Wilkinson. "Bactericidal Activity of Different Honeys against Pathogenic Bacteria." Archives of Medical Research. U.S. National Library of Medicine, 2005. NCBI.nlm.nih.gov/pubmed/16099322.

Steckelberg, MD, James M. "Honey: An Effective Cough Remedy?" Mayo Clinic. Mayo Foundation for Medical Education and Research. May 2, 2018. MayoClinic.org/diseases-conditions/common-cold/expert-answers/honey/faq-20058031.

Chapter Two
Burns, David. "Beekeeping Success Depends on Your Queen." YouTube video, 1:18. March 11, 2020. youtu.be/u7INchQS_pY.

Chapter Four
"Federal and State Bee Laws and Regulations." Your online apiary resource for all things beekeeping. Accessed March 11, 2020. Beesource.com/resources/usda/federal-and-state-bee-laws-and-regulations.

Chapter Eight
"Mid Atlantic Apiculture Research & Extension Consortium." MAAREC. Accessed March 11, 2020. MAAREC.cas.psu.edu.

Chapter Nine
"Guidance For Industry Food Labeling Guide." Center for Food Safety and Applied Nutrition. U.S. Food and Drug Administration. Accessed March 11, 2020. FDA.gov/regulatory-information/search-fda-guidance-documents/guidance-industry-food-labeling-guide.

INDEX

A

American foulbrood, 80, 97
apiary, 59, 89, 96

B

balms and salves, 138
bearding, 90
bee bread, 23
bee feeder, 58, 69–70, 91
bee feeding, 61, 71, 91, 93, 103, 109
beekeeper
  associations/clubs, 1, 8, 12,
    13–14, 111
  backyard, 4, 7, 11, 15, 41, 48,
    58, 109–110
  commercial, 6, 15, 48
  urban, 12, 50
beekeeping
  considerations, 13–15
  disease/pest control, 15, 96–97
  first steps, 4, 8–9
  history of, 7
  legal considerations, 10, 14, 58
  medications, 15–16
  natural, 15, 16, 17
  nectar sources, 10–11, 16, 71, 91.
    See also nectar
  organic, 15, 16–17, 116, 138
bees
  buying, 30, 102
  drones, 20, 22, 24, 26, 31–32
  feral (wild), 13, 14, 61, 100
  foragers, 21, 28–29, 45, 73, 88–89
  lifecycle of, 22–23

queen, 24–26, 69–70.
    See also queens
  scout, 21, 91, 100, 142
  types/characteristics, 33
  worker, 20, 22, 23, 24, 26–29, 48
beeswax, 116–117, 137–147
beeswax candles, 140–141
beeswax food wraps, 142
brood, 22
  area, 7, 80
  box, 38, 42, 58, 103
  cells, 78
  comb, 10, 38
  disease, foulbrood, 15, 80, 97
  frame, 80, 89
  nest, 22, 26, 28, 39, 83
  pheromones, 23
  winter, 91

C

candy board, 71, 91, 93, 103
colonies
  dividing, 98
  drone production, 32
  inspecting, 48, 96
  mite treatment, 17
  swarms, spring collecting, 89
  wintering, 93
colony collapse disorder
  (CCD), 6, 7
comb. *See also* propolis
  aging, 80
  building, 28
  handling of, 111, 113
  health of, 16, 96–97

ACKNOWLEDGMENTS

Writing a book is fun, but it's not possible without years of study, work, experimentation, and mentoring by others. We are thankful for the mentors in our life—there are too many to list. One who cannot go without mention is our good friend and supporter Jon Zawislak, who David met while he was being certified as an EAS Master Beekeeper; Jon continues to be a collaborator on podcasts (*Hive Talk*), videos, and writing projects. We are grateful for the time and effort given to us by so many people, including our editing assistant, Jennifer Copass, who gave fantastic guidance and suggestions throughout this project.

ABOUT THE AUTHORS

 **DAVID AND SHERI BURNS** are beekeepers with over 29 years of experience in hobby beekeeping. Sheri has a BA in leadership and management. She operates the business side of their online company, HoneybeesOnline.com, and supervises their farm store, which sells beekeeping supplies and honey products. David, a graduate of Lincoln Christian University, is a lifelong educator who received his Certified Master Beekeeper designation from the Eastern Apicultural Society in 2010. Sheri and David are parents to six children. David is also a competitive sportsman, and Sheri invests her time in a vast array of hobbies.

Printed in the USA
CPSIA information can be obtained
at www.ICGtesting.com
CBHW071603160224
4317CB00003BA/17

9 781647 395148